THE ROYAL INSTITUTE OF NAVIGATION

THE USE OF
RADAR AT SEA

Edited by
Captain F. J. Wylie
O.B.E., F.R.I.N., F.I.E.R.E., R.N. (ret)

Foreword to the first edition by
Sir Robert Watson-Watt C.B., F.R.S.

HOLLIS & CARTER
LONDON SYDNEY
TORONTO

British Library Cataloguing
in Publication Data
Royal Institute of Navigation
The use of radar at sea. – 5th revised ed
1. Radar in navigation
I. Title II. Wylie, Francis James
623.89'33   VK560
ISBN 0-370-30091-2

All rights reserved
This edition © The Royal Institute of Navigation 1978
Printed and bound in Great Britain for
The Royal Institute of Navigation
1 Kensington Gore, London, SW7 2AT
and
Hollis & Carter
an associate company of The Bodley Head Ltd
9 Bow Street, London, WC2E 7AL
by Cox & Wyman Ltd, Fakenham
Set in Monotype Imprint
First published 1952
Second edition 1954
Third, revised, edition 1958
Fourth, revised, edition 1968
Fifth, revised, edition 1978

THE USE OF RADAR AT SEA

FOREWORD
TO THE FIRST EDITION

by SIR ROBERT WATSON-WATT, C.B., F.R.S.

BRITISH radar was born just seventeen years ago. Although it has been described as the best advertised secret of all time, its first ten years were, in fact, wrapped in a secrecy that was astonishingly complete. Two more years passed before it was effectively introduced (necessarily on a very limited scale) into civil operation. It is not surprising that the last five of this seventeen-year life have not sufficed for the propagation of a generally available body of doctrine on how current radar devices can best be used in seagoing practice. Indeed the doctrine is still being established and extended, and its propagation must be a tentative and intermittent process. Of the now numerous books about radar, only a very few have yet attempted to carry to the marine user a summary of those lessons which have already been learnt, or a suggestion of how the succeeding lessons can most quickly and painlessly be learnt.

This civil task is intrinsically more difficult than was the corresponding military task of the war years. In the armed forces and their supply services the provider and the user could be brought into more intimate and continuing contact than can user and provider in the diversified arts of peace. The user had recognized and substantially effective channels through which he could insist on the meeting, by the provider, of what he himself supposed to be his principal needs. He might also be wrong in his judgment; he was indeed often wrong, largely because Mark II had to go into production before he had sufficient experience in the use of Mark I to show what Mark II should have been like. But the military user had the further advantage that the elastic economics, which were dictated by the grim necessities of war, allowed the early appearance of a Mark III free from some, at least, of the defects that should have been absent from Mark II. The wide terraces of civil practice must, of commercial necessity, be fewer and more steeply separated than the numerous shallow steps of the staircase of specialized military provision.

The Institute of Navigation was conceived as a contribution to the better exchange of experience and opinion between users and providers. The civil user of radar had already, before its foundation, enjoyed very great advantages from military radar experience; these advantages have been increased by the work of the Institute and in particular by its periodical publications.

Now the Institute is enabled, thanks to the unique, and uniquely diversified, reservoirs of specialized knowledge, pioneering zeal, and tireless voluntary exertion on which it can draw, to offer to the marine radar user an interim guide to the art of extracting the maximum of dependable information from present-day shipborne and shore-based radar. Its limitations and defects will be clearer to its contributors and editors than to other readers, and it is for that reason that the word 'interim' is advisably applied to this first edition. They have learnt again the great lessons of all teaching, that the teacher learns from his own effort to teach; that the student must teach his instructor if education is to advance; that failures teach more than do successes.

This first cooperative manual of *The Use of Radar at Sea* is a work confidently offered by the Institute as a much-needed immediate aid to the marine radar user; it is confidently offered, too, for that most stringent criticism by the experienced user and by the earnest student which alone can ensure the still greater utility of an early second edition.

The Institute is profoundly grateful to those who have provided and arranged the text and the copious and indispensable illustrations of this manual. Authors, editors and Institute alike will measure their success by the constructive criticism which it must elicit from those who have in their hands the future of marine radar, the seagoing users and the largely shore-bound provider. And of these two it is the user who must be the more vocal, if only because, difficult as it may be to produce a Mark IV radar, that process is much easier and quicker than that of evolving a Mark IV user who would conform to the radar designer's abstract and de-humanized blue-print of an ideal marine radar user!

ROBERT WATSON-WATT
London, 1952

PREFACE
TO THE FIRST EDITION

THIS book has been produced by the Institute of Navigation to meet the need for a comprehensive treatment of the use of radar at sea. It is intended primarily to help the seagoing user and to serve as a manual of instruction in navigation schools. It should, however, be of assistance to many others who are concerned with marine radar in different and, perhaps, less direct ways.

The aims of the production have been to give the mariner a complete and practical guide to the applications of marine radar and to the many factors which affect the results obtainable, and to provide a book of reference in which most of what he will want to know can readily be found. The preparation of such a volume, which has to deal with complex matters in a comparatively simple and yet authoritative manner, has naturally presented many problems.

When the book was first discussed, it was agreed that the only way to ensure a particular aspect of the subject being treated entirely competently was to invite a contribution from an acknowledged expert in that field. The direct consequence of this was the necessity to dovetail the various contributions into the overall plan and, as far as possible, to put them into a common language. The last of these aims was no easy one, and some indulgence on the reader's part may be claimed if there are irregularities in style.

It is at least hoped that a uniform and palatable level of approach to the subject has been achieved and that the reader will find it a comparatively simple business to follow the theme through the various stages of its development. Some apologies, or at least acknowledgment, to the authors might be made at this point, since they have had to endure the alteration of what, to them, must have been perfectly satisfactory statements on their subjects, to conform with a pattern of which they were never fully aware.

Information which has helped in the book's preparation has come from a wide variety of sources. Many individuals have contributed by their work and interest; radar manufacturers, both here and abroad, navigation schools, shipping companies and a great many seagoing officers of the Royal and Merchant Navies have helped by giving the benefit of their knowledge and experience. Particular mention should be made of the National Physical Laboratory and the Meteorological Office for their help with Chapter 5; H.M. School of Navigation for its help and advice on the navigational aspects of the text; Trinity House for its useful comments on the text; the Ministry of Transport for the assistance of their Operational Research Group and others, and the Chamber of Shipping and Liverpool Steam Ship Owners' Association for the work done by their Radio Advisory Service, and in particular by Mr. T. W. Welch.

1952   F. J. W.

PREFACE
TO THE FIFTH (REVISED) EDITION

This book was first issued in the Coronation year of Her Majesty the Queen and we, therefore, have the honour and privilege of sharing the Silver Jubilee year with her. Twenty-five years is a longish time for a semi-technical book to remain in demand, particularly when it deals with such a fast-moving subject. This is its third major revision, the last being done ten years ago.

The only chapters which remain relatively untouched are 4, 5, 6 and 7. These cover the basic physics of the subject as far as the user is concerned. No doubt knowledge of the subject has broadened but not in a way to invalidate those chapters. Chapters 1 to 3 and 8 and 13 to 15 all needed bringing up to date, mostly in minor ways, the continued supersession of the valve by solid-state devices, improvements to charting methods and the introduction of the Hertz as the unit of frequency being typical of the decade's changes.

In Chapter 10, Radar and the Rule of the Road, major changes were necessitated by the introduction in July 1977 of the Regulations for Preventing Collisions at Sea 1972. Much of Chapter 11 had to go because of developments in the field of buoyage and passive reflectors and racons. Similarly, large changes in the equipment of Shore Radar Stations, particularly radar displays, required changes in Chapter 12. However, the advances most nearly affecting the shipborne user have been in radar plotting. The introduction of fully automatic radar plotters, which purport to do all the work for the radar observer except to appraise the displayed situation, has brought with it competition and controversy. It was decided to leave Chapter 9 to stand almost untouched as a statement of the basic ideas and methods of plotting and to give a much deeper study and a description of the new equipments in an Annex. As Chapter 16 appeared to be somewhat out of date and as most manufacturers supply excellent technical manuals for operators, it was decided to cancel it.

A number of people have been extremely kind in helping me with this revision. I have had the latest information in the form of Brochures and Manuals from I.B.M. (MABS), Iotron (Digiplot), Norcontrol (Databridge) and Sperry Marine Systems (CA system), from Marconi International Marine (Predictor), Decca Radar (Clearscan and AC), Kelvin Hughes (Situation Display), Barr & Stroud and Commander Parrish (Autoplot), Raytheon Copenhagen (TM/CA). For all of this I am very grateful indeed. On a more personal note, I am greatly indebted to Mr. J. F. Tyler (Marconi Marine) for doing two chapters and an Appendix and many useful comments; to Captain I. S. Mackay R.N. for Chapter 8; to Mr. A. P. Tuthill for script and photographs for Chapter 12, with Commander J. F. Knight R.N. (Liverpool), Captain T. Hand (Tees) and Captain R. K. Emden D.S.C., R.N. (C.N.I.S.) on the operational side; to Mr. E. R.

Richards (Trinity House) for doing Chapter 11; to Messrs. J. C. Herther and J. S. Coolbaugh (Iotron) for much help of all kinds and particularly, of course, on the Automatics; to Mr. J. H. Beattie (Decca Radar) also for much general help. I must also add my thanks to the nine anonymous (to me), independent inquisitors who dealt very kindly but thoroughly (!) with my draft of the Annex. Even more to Michael Richey and Frank George who insulated me from the inquisition and, somehow, dovetailed their work into mine.

November 1977   F. J. W.

CONTENTS

FOREWORD, v

PREFACES, vii

Chapter 1. RADAR PRINCIPLES AND GENERAL CHARACTERISTICS

The echo principle – Brief history – The radio wave – The propagation of waves – The display – Resolution or discrimination – Radar equipment. 1

Chapter 2. THE RADAR EQUIPMENT

The transmitter – The aerial – Transmit-receive arrangements – The receiver – The display or indicator – True and relative presentation – The cathode-ray tube – The performance monitor. 22

Chapter 3. OPERATIONAL CONTROLS

The controls – Adjustment of the display – Gain and anti-clutter – Tuning – The performance monitor – Measurement of bearing – Measurement of range – Routine check procedure. 46

Chapter 4. PROPAGATION OF WAVES AND RESPONSE OF TARGETS

Transmitter signal strength and range – Attenuation and shadowing – Echo strength and range – Radiation and coverage diagrams – Horizontal diagrams – Vertical coverage – Aerial height and wavelength – Range and height of first detection – Target size and shape – Equivalent echoing area – The effect of target position – Response and behaviour of practical targets – Typical performance data. 56

Chapter 5. RADAR METEOROLOGY

Standard propagation – Non-standard propagation – Sub-refraction – Super-refraction – Attenuation by atmospheric gases – Attenuation by fog, rain, hail and snow. 88

Chapter 6. INTERPRETATION OF THE DISPLAY

Echoes from land – Echoes from ice – Echoes from the sea – Echoes from clouds – Artificial targets on land – Echoes from small isolated targets – Echoes from ships – Distinguishing between targets – Appreciation of movement, true and relative. 98

Chapter 7. UNWANTED ECHOES AND EFFECTS

Sea clutter – Weather effects – Anomalous propagation – Second-trace echoes – Multiple echoes – Interference – Shadow areas – False or indirect echoes – Side-lobe effects – Set faults.   121

Chapter 8. RADAR AS AN AID TO NAVIGATION

Landfall – Long-range position finding – Coastal navigation – Radar range and visual bearing – Radar ranges as position circles – Radar range and radar bearing – Radar range as a clearing line – Coasting in general – Pilotage – Orientation of the display – Buoys and buoyed channels – Other ships – Anchoring – The operation of the radar set – Errors – Shipboard aids to the use of radar – Radar charts.   144

Chapter 9. RADAR FOR COLLISION AVOIDANCE

Visual and radar observation compared – The information required – The true plot – The relative plot – Speed-vector diagram – Targets to be plotted – Occasions for use in open waters – Use in pilotage waters – Reporting from the plot – Errors.   170

Chapter 10. RADAR AND THE RULE OF THE ROAD AT SEA

Radar detection versus sighting – Radar and 'safe speed' – Ascertaining the position of a ship by radar – Radar brings responsibility.   191

Chapter 11. AIDS TO INCREASING ECHO STRENGTH AND TO IDENTIFICATION

Factors affecting the use and selection of reflectors – Radar reflector identification – Radar beacons, in-band and cross-band.   196

Chapter 12. SHORE-BASED RADAR

The functions of shore-based radar – The operation of shore-based radar – The Liverpool shore radar station – Teesport radar.   219

Chapter 13. THE IMPORTANCE OF RECORDING EXPERIENCE

Log of targets – Operational log – Permanent information.   234

Chapter 14. SIMPLE MAINTENANCE

Routine maintenance of moving parts – Routine electrical maintenance – Performance checking – Fault finding – Location of faults – A fault list.   242

Chapter 15. THE RADAR-EFFICIENT SHIP

Siting the scanner – Siting the display – Siting the transmitter – Interference – Operation – Maintenance. 252

Chapter 16. RADAR IN THE FUTURE

The reliability of sets – Consistency of performance – Detection range and target identification – Improvement of resolution – Improvement in PPI presentation – Improvement in aids to the use of radar – Radar for avoiding collision – Radar as an aid to berthing – Radar in fog. 258

Annex. COLLISION AVOIDANCE SYSTEMS 271

The radar plot – Plotting systems – Automatic and computerized systems – Advantages and limitations of automatic equipment.

APPENDICES

I	Echo Recognition Table	286
II	Marine Radar Performance Specification, 1968	292
III	Some Constants, Formulae and Useful Data	299
IV	Useful Test Equipment	302
V	A Short Glossary of Terms	305
VI	Merchant Shipping (Radar) Rules 1976 – Provision	309
VII	Merchant Shipping Notice M784 – The use of radar	311
VIII	Symbols for Controls on Marine Navigational Radar Equipment	314

INDEX, 323

PLATES

Plate

1a	Sectional view of the cavity magnetron.	27
1b	The magnetron in position.	28
2a	Two types of scanner.	29
2b	Slotted waveguide aerial.	29
3	8-mm. radar at London Airport.	30
4	Use of differentiator to reduce rain clutter.	31
5	Typical display units.	32
6	PPI display showing range rings and heading marker.	33
7	PPI display in good adjustment.	47
8	A typical PPI picture.	100
9	Distant land.	101
10	Steep cliffs.	101
11	Sloping rock face.	102
12	Low-lying foreshore.	102
13	Change of appearance with angle of view.	103
14	Shielding.	104
15	Land and ice.	105
16	Ice echoes.	105
17	Various types of echo.	106
18	A built-up area.	106
19	Clearing the picture by reducing gain.	115
20	Bridges in the Thames.	116
21	Ships at medium range.	116
22	Ships at close range.	117
23	Buoy echoes.	118
24	Wake of a ship depicted by the smoothing of clutter.	118
25	Sea clutter.	127
26	Lines of rollers.	127
27	Light to moderate rain squalls.	128
28	Heavy rain squalls.	128
29	Multiple echoes.	129
30	Radar interference.	129
31	Shadow sectors.	138
32	Indirect echoes.	138
33	False echoes.	139
34	Side-lobe echoes.	139
35	Spoking.	140
36	Fix by radar range and visual bearing.	149
37	Fix by radar range circles.	150
38	Estimating the bearing of an isolated target.	150
39	A record of radar fixing.	151
40	A turn with heading-upward presentation.	160

	PLATES	XV
41	A turn with north-upward presentation.	160
42	Use of the variable range marker.	161
43	Compass-datum plotting device.	178
44	A collapsible corner reflector hoisted in a ship's boat.	203
45	A pentagonal cluster mounted on a buoy.	204
46	Signals from two experimental ramarks.	204
47	In-band racon signal.	205
48	Liverpool Bay, overall PPI picture.	225
49	Liverpool radar operations room.	226
50	Liverpool tower.	227
51	Dover Strait from St. Margaret's Bay.	228
52	Digiplot.	229

Illustrations have been provided by the following, to whom grateful acknowledgment is made:

>The Admiralty Signal and Radar Establishment
>Associated Electrical Industries Ltd.
>The Corporation of Trinity House
>Decca Radar Ltd.
>The Dubilier Condenser Co. Ltd.
>Mr. G. Hermansen
>H.M. Stationery Office
>Iotron Corporation
>Kelvin & Hughes
>Marconi International Marine Co. Ltd.
>Mersey Docks & Harbour Company
>The Ministry of Transport & Civil Aviation
>The Sperry Gyroscope Co. Ltd.
>The Swedish Defence Research Board

ORIGINAL CONTRIBUTORS

N. Bell (*Trinity House*)

D. J. Cashmore (*Ministry of Transport*)

Captain W. R. Colbeck, R.N.R. (*Mersey Docks & Harbour Board*)

R. F. Hansford (*Decca Radar Ltd.*)

L. Harris-Ward (*Ministry of Transport*)

H. E. Hogben (*Admiralty Signal & Radar Establishment*)

M. W. Kaye (*Radio Advisory Service*)

L. S. Le Page (*Ministry of Transport*)

A. L. P. Milwright (*Admiralty Signal & Radar Establishment*)

Commander H. D. Stevenson, R.A.N. (*H.M. School of Navigation & Direction*)

J. R. Webster (*Ministry of Transport*)

T. W. Welch (*Radio Advisory Service*)

Captain F. J. Wylie, O.B.E., R.N.(ret.) (*Radio Advisory Service*)

1

RADAR PRINCIPLES AND GENERAL CHARACTERISTICS

RADAR is a method of detecting objects and ascertaining their range and bearing. The captain of a ship requires to know at all times his position in relation to nearby land, and the position of other ships in the vicinity in relation to his own. Radar can give him this information quickly and accurately, and in any sort of visibility. The picture that radar presents, however, will not be as clear or as comprehensive as that presented by the eye in clear weather; there is in fact 'more to it than meets the eye.' For this reason it requires a skilled and careful interpretation, an understanding of what in fact is being presented. A knowledge of the principles and of the characteristics and performance of radar is therefore necessary if the best value is to be obtained from the service that it offers. To understand fully what is happening it is necessary to know why it should happen. This book therefore endeavours to deal faithfully with the theory and practice of marine radar, and to lead the reader from an understanding of its basic principles through every aspect of its use as an aid to navigation.

The first chapter describes the basic principles and the historical background of radar and then examines the characteristics of radio waves and of the radar equipment itself.

The echo principle
The word Radar was coined from the phrase Radio Detection and Ranging. In Britain, for security reasons, the original term for it was R.D.F., or Radio Direction Finding. Essentially, as these terms imply, it is a system for determining the position of objects by echo-range and bearing.

The principle of echo-ranging is by no means new. Many a ship has used it when closing the land in fog to obtain a distance from a cliff or headland. If the ship's syren or whistle is sounded briefly, after an interval the echo from the cliff will return. A stop watch, started at the commencement of the blast and stopped at the beginning of the echo, gives the time taken for the sound to travel to the cliff and back. The speed of sound waves in still air is known to be about 1100 feet per second, and therefore if the time interval is found to be, say, 6 seconds, the ship must be about 3300 feet from the cliff.

The same principle is used in the echo sounder, in which a short burst or pulse of sound waves, usually of supersonic frequency, is transmitted downwards through the water and an echo is received from the sea bottom. The speed of sound waves in water is known and the echo time interval is measured automatically. A device which employs the same principle but is more akin to radar is the anti-submarine Asdic. In this case the

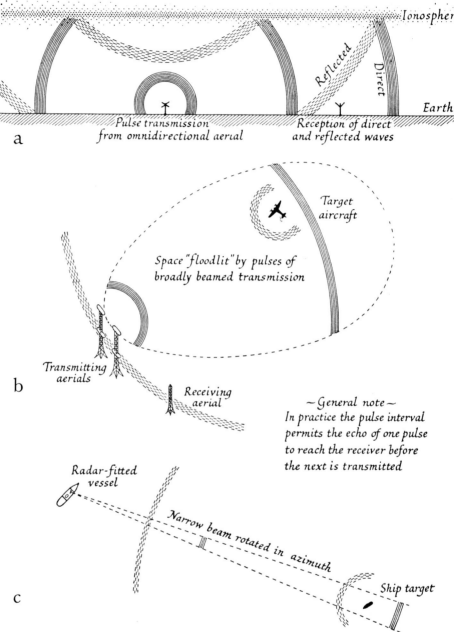

(a) Ionospheric research 1925. Height of reflecting layer is calculated from arrival times of direct and reflected waves
(b) Early air defence radar 1935. Receiving equipment obtains range, azimuth and elevation
(c) Marine radar 1946 et seq. Receiver indicates range and bearing

Fig. 1. The evolution of radar.

supersonic beam is used to search for targets under water. The direction of the beam when an echo is received is accurately known and hence the direction of the target, as well as its range, is given.

To be of practical value for marine navigation, any device of this sort must offer high accuracy and long-range working. Ranges of large ship targets at 10 to 15 miles and of land at perhaps twice these distances are needed. For these purposes sound waves are useless, as the effect of wind on the range of fog signals and the difficulty of determining the direction of the source clearly demonstrate. The fallibility of sound locators for detecting the approach of aircraft will also be remembered. A system which will fulfil marine navigational requirements must be largely independent of visibility and weather and be capable of detecting and fixing the positions of quite small targets.

Brief historical survey

The problem which, in the 1930's, confronted those in Britain who were responsible for defence against air attack was precisely that of devising a system more effective than sound location. As a carrier of energy, radio waves seemed to be the obvious choice, since they satisfy the low visibility requirement and can be sent in desired directions, albeit in somewhat wide beams. But extremely little was known in those days of the echoing properties of targets in relation to radio waves.

The evolution of modern marine radar may be said to have begun in the middle 1920's in the work of scientists, notably in Britain and the U.S.A., who were engaged in determining the heights above the Earth of the electrically conductive layers of the atmosphere which reflect radio waves (Fig. 1). Prominent in this work in Britain was the late Sir Edward Appleton with whom were associated M. A. F. Barnett and the late R. A. (Sir Robert) Watson-Watt.

The possibility of obtaining reflection from targets, and so determining their positions was foreseen at this time, and as early as 1922 Marconi is known to have referred to it. Some twelve years elapsed before the problem approached a practical solution, although evidence was accumulating that the presence of even quite small objects in the path of a radio wave could be detected by the energy reflected from them. Meanwhile, equipment suitable for the transmission and reception of pulse-modulated radio waves and for the measurement of the minute time intervals involved was under continuous development.

In 1935 when the need for adequate warning of the approach and movements of hostile aircraft was exercising the minds of many responsible persons in Britain, a radio solution to the problem was little more than a hope. It seemed however the only hope and, as a result, work on the design and erection of experimental equipment was given very high priority and put under the control of Watson-Watt at Bawdsey on the Suffolk coast. The early trials gave promise which exceeded the expectations even of those conducting them and led, just in time, to a system of radar detection which was of incalculable value in the defence of the country.

Naval authorities were no less concerned to possess equipment which would give their vessels the ability to detect aircraft in the air and ships at sea, particularly in fog and at night. The first Bawdsey equipment was bulky and required high aerial towers and the main effort of the naval designers in Britain was directed towards producing equipment small enough for ships to accommodate and aerials which could be mounted on their masts. The pressure put into this work resulted in the first prototype equipments being fitted in the battleship *Rodney* and the cruiser *Sheffield* in 1939. So radar went afloat.

The outbreak of war, coupled with these early successes, led to a tremendous increase in the effort to meet all requirements. As radar developed, its possible usefulness became more evident; new vistas opened and new urgencies came into being. The need to detect low-flying aircraft and small surface targets, together with the demand for a bearing accuracy greater than that obtainable with the early equipment, forced the designers to shorter and shorter wavelengths; as is described later, these permit the concentration of the radio energy into a narrow beam and keep a substantial part of it close to the surface of the sea. The need for shortening the wavelength to a few centimetres had been apparent some time before the inherent technical problems could be solved. No valve was available which, on such wavelengths, could develop the power necessary to obtain a perceptible echo from small targets. A stage had been reached where the physical dimensions of conventional valves were comparable with the required wavelength, and apparently insurmountable difficulties were encountered.

However, at this critical moment, the work of Randall and Boot at Birmingham University resulted in 1940 in the development of a valve of revolutionary design, the cavity magnetron. It was a revolution not only of valve design but of radar technique, and it made possible from 1943 onwards the detection at close and medium range of objects on the surface as small as a submarine's periscope. The accuracy of resulting equipment led to it becoming a *sine qua non* in the hunting of U-boats and in the control of gunfire. As an earlier type of naval radar had secured the detection of the *Bismarck* in the Denmark Strait in 1941, and thus led to her destruction, so the later devices helped to break the submarine menace and directly contributed in 1944 to the sinking of the *Scharnhorst* in darkness at a range of about 6 miles.

These equipments were the rugged ancestors of the sleek but effective marine radar of today with which this book is concerned.

FUNDAMENTALS OF THE USE OF RADAR

To appreciate the capabilities and limitations of radar, and so to be able to use it to full advantage, it is necessary to understand the character and

RADAR PRINCIPLES AND GENERAL CHARACTERISTICS

behaviour of radio waves and to appreciate the principles of their generation and reception, including the presentation of the echo to the observer. In many respects this knowledge need only be superficial, but neglect of the subject would lead to imperfection in the important art of interpreting the radar picture.

Essentially, a marine radar equipment consists of a means of generating radio-frequency oscillations in a particular form; sending them out into space as a narrow beam of radio waves, which is rotated continuously in azimuth; receiving the echoes returned from any target in the area surrounding the ship; and displaying the returned echoes visually on a screen in such a way that the bearing and range of each target from the ship are immediately evident. From this brief description it will be seen that the main parts of a radar are a transmitter, an aerial, a receiver and a display or indicator unit. The functions of these and the principal units they contain are described in the next chapter. The remainder of the present chapter is devoted to examining the character and behaviour of radio waves and the effect of these on the picture which radar can display to the observer.

The radio wave

Radio waves have characteristics common to other forms of wave motion so that it is permissible to explain them by comparison with, for example, ocean waves. In those it is obvious that, to an approximation, wave motion consists of a succession of crests and troughs which follow one another at equal intervals and move along at a constant speed. A piece of driftwood will be seen to rise and fall as the waves pass, but, in the absence of wind or current, not to progress in any direction. This shows that the wave, caused originally by some remote disturbance, progresses through the medium (in this case the ocean) at a constant speed but that the medium itself is not permanently displaced. It will not be necessary to quote evidence to show that the wave contains energy.

The distance between successive crests is the *wavelength* (symbol λ, the Greek lambda). A complete alternation or oscillation from one crest through a trough to the next crest is called a *cycle*. The number of complete waves which pass a fixed point in a given time is the *frequency* (symbol f), formerly expressed in cycles per second (c/s) and now called Hertz (Hz). Obviously the number of Hertz depends upon the wavelength and the speed at which the wave is moving (symbol c) (Fig. 2). A wave of 2-metre wavelength moving at a speed of 10 metres per second must oscillate at a frequency of 5 Hertz. So the expression connecting these factors is

Speed = Frequency × Wavelength; or $c = f \times \lambda$.

One further fact can be usefully demonstrated by the analogy with waves in water. A stone dropped into a pond will show how wave motion moves outwards at an equal speed in all directions unless it is obstructed.

Radio waves have all the characteristics that have been mentioned. The essential difference is that they are electromagnetic waves and are caused by electromagnetic instead of mechanical disturbances. Radio waves belong to the same family as light waves though their wavelengths are much longer, which factor greatly affects their characteristic behaviour. When speaking of them electrical rather than magnetic terms are used, their strength being measured in volts and their power in watts. The energy in them is made apparent by the fact that they induce electrical currents in materials with which they come into contact.

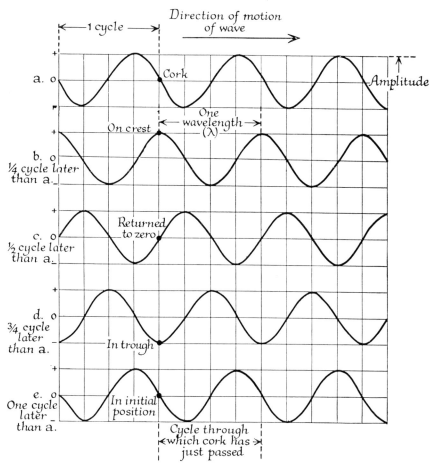

Fig. 2. *Wave motion.*

An important characteristic of radio waves in connection with radar is their polarization. A radio wave has an electrical axis and a magnetic axis which are at right angles to one another. The orientation in space of these axes is called polarization and it is described in terms of the direction of the electrical axis. Horizontal polarization for radar waves is the more

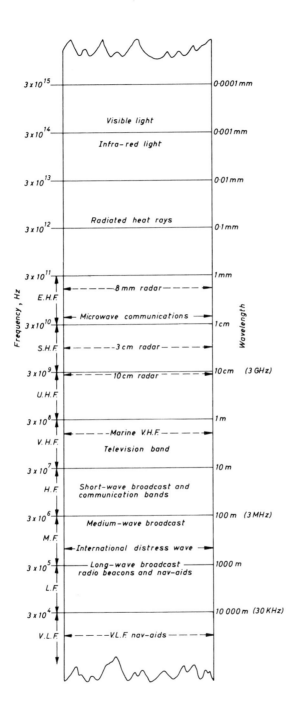

Fig. 3.

usual as it is considered to produce better echoes from the majority of maritime targets. However, vertical polarization may be employed. Further, the aerial can be so constructed that the axes of the radiated wave rotate and the wave proceeds forward in a spiral fashion; this is known as circular polarization. It is sometimes used as a means of reducing the echoes from rain, &c. (see page 25 and elsewhere).

The wavelengths of radio waves which are at present usable lie between about 20,000 metres and 4 mm. (Fig. 3). Their speed in free space is constant at about 300 million metres or 161,800 nautical miles per second, and this value is commonly used when considering their passage through the atmosphere. As the speed is constant, when the frequency is increased, the wavelength will decrease. An example will show how the formula quoted is used to translate wavelength into frequency.

To find the frequency of a radio station transmitting on a wavelength of 1500 metres:

$$f = \frac{c}{\lambda}, \text{ or Frequency} = \frac{300,000,000}{1500} = 200,000 \text{ Hertz}.$$

Radio frequencies are usually expressed in terms of kiloHertz (kHz), MegaHertz (MHz) and GigaHertz (GHz), denoting respectively thousands and millions of Hertz and thousands of MegaHertz. For very short time intervals, microseconds (millionths of a second) are used. The wavelengths commonly used for marine radar are about 10 cm. (3000 MHz) or 3 cm. (10,000 MHz). Radio waves of this order of wavelength are called microwaves.

The general nature of radio waves can be visualized from the simple analogies given. Their behaviour depends very much on their wavelength and on the manner in which they are radiated into space. This subject is fully dealt with in Chapter 4, but it should be mentioned here that they may be beamed in a particular direction and may suffer reflection, refraction and diffraction, much as light waves do, depending upon the nature of the media through which they pass and the objects with which they come into contact.

Directivity of the transmitted wave

Disregarding for the moment how the radio wave is produced, the next step is to consider the problem of starting the wave on its journey into space. Here again a direct comparison with light waves is useful. A small electric bulb suspended in the centre of a dark space will give a small amount of illumination in all directions. If this same bulb is placed at the focus of the reflector of a motor-car headlamp it will send out a narrow beam of great intensity. Radio waves may be focused in the same way and, from the radar point of view, this has two advantages: the direction of transmission can be accurately known and the power sent in the required direction (and hence the likelihood of obtaining an echo) is enormously increased. So far as strength and directivity are concerned, the lighthouse is an excellent analogy.

The size of the reflector needed to concentrate electromagnetic waves into a beam of given width depends upon the wavelength used; the longer the wave the larger must be the reflector. Hence, when employing a reflector of dimensions suitable for ships, a very short wavelength must be used to obtain a narrow beam. This is one of the reasons for using microwaves for marine radar. On a wavelength of 3 cm. a beam-width of about $1\frac{1}{2}°$ can be obtained from a reflector 5 feet wide or a 0·75° beam from a 10-foot reflector.

Accuracy in measuring direction is needed only in the horizontal plane, that is, in azimuth. In the vertical plane the beam is made wider, so that rolling of the ship will not cause it to miss targets. It will be realized that the wider the beam in any direction the less will be its intensity. This is therefore a matter for compromise and it will be found that some radars use as much as 30° vertical beam-width and others as little as 15°.

As the beam is narrow it must be rotated in azimuth so that radar information may be obtained from all directions round the ship. In common with other forms of radio, the device used for sending the waves off into space is known as the *aerial* or *antenna*. Since in this case it rotates and scans the surrounding area it is frequently called the *scanner*.

A reflector which sends out, in a divergent beam of a certain angle, the energy which is directed at it from its focal point will also concentrate at the same point any energy from external sources which reaches it from within the same angle. That is to say the aerial is as directive for reception as it is for transmission. This not only favours accuracy of bearing measurement but also gives a gain in the intensity of the received wave.

Radar pulse-length and repetition frequency

Hitherto in considering radio waves the reader has been allowed to imagine a continuously generated wave motion without beginning or end. From the mention of echo-ranging early in the chapter, however, it will be appreciated that the returning echo will resemble the outgoing wave fairly closely and that accurate measurement of the time interval between them will require the former to be interrupted at intervals and to be suitably shaped.

A primitive experiment in echo-ranging by the human voice will show that the best results are obtained if the transmission consists of a short sharp sound. The shortness ensures that the transmission will have ceased before the echo returns and the sharpness, or the sudden rise to full power, will make the instant of the echo's return immediately perceptible. The sound wave consists, therefore, of a few oscillations at maximum power. Similarly the radar transmitter is required to produce a short *pulse* of oscillations, rising rapidly to a maximum strength or amplitude which it maintains until it is rapidly cut off. The outline shape (envelope) of such a pulse would be somewhat as shown in Fig. 4 at (*b*) which gives rise to the description *square pulse*.

There are two factors here which affect the design of radar equipment: the duration of the pulses, or *pulse-length*, and the interval between them. The latter is usually defined by the number of pulses per second, called the *pulse repetition frequency* (p.r.f.) (Fig. 4 at (*c*)). If it is desired to receive

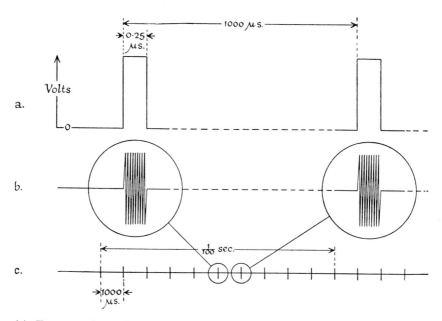

(*a*) *Two successive modulator (d.c.) pulses as applied to magnetron*
(*b*) *The resulting pulses of radio-frequency energy*
(*c*) *A succession of such pulses as transmitted: the number of pulses transmitted per second = p.r.f. (in the example, 1000 pulse/sec)*

Fig. 4. *d.c. and r.f. pulses and pulse repetition frequency.*

echoes from targets as close as 50 yards from the transmitter, the pulse will have to be cut off before the beginning of the wave has had time to travel to the target and back: a total distance of 100 yards. A radio wave travels 328 yards in 1 microsecond, so that the pulse, in this case, must not be longer than 0·3 microseconds (Fig. 5). If it is also desired to receive echoes from targets up to say, 30 miles' range, the interval between pulses must be long enough to enable the wave to travel twice this distance, i.e. 370·4 microseconds. This gives a maximum p.r.f. of 2700 pulses per second.

It will be noticed that in the case mentioned the transmitter is required to oscillate for 0·3 microseconds and then to rest for 370·1 microseconds. Because of this a small valve may be used to generate very high power since it has relatively long intervals in which to cool. In practice the p.r.f. is usually between 500 and 2000 pulses per second. In these circumstances a valve no bigger than a 25-watt lamp will give a peak power of 60 kW.

The propagation of waves

A little must be said here of the passage of radio waves through space after they leave the aerial and of the echoing properties of targets, although these subjects are dealt with more fully in Chapter 4. If the similarity between radio waves and light waves is borne in mind there will be little difficulty in appreciating their behaviour.

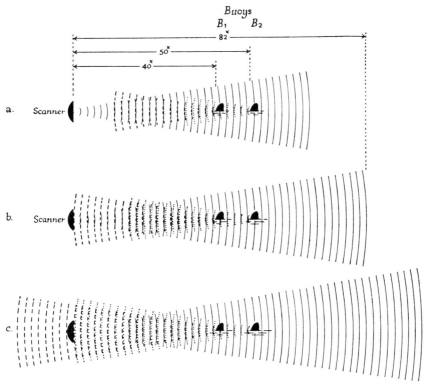

(a) Radar pulse being radiated. Echoes from both buoys returning
(b) Echo from B_1 has reached scanner just before transmission has ceased
(c) Transmission has ceased. Echo from B_2 reaches scanner
Pulse-length 0.25μ sec (82 yards): minimum range 41 yards

Fig. 5. Minimum range.

Refraction in the atmosphere. It is well known that, to an extent depending on its angle of incidence, a ray of light in passing from one medium to another of different density may be bent or refracted from its path. This occurs when the speed of the wave is different in the two media. The difference between the speed in a particular medium and that in free space is usually expressed as a ratio, known as the refractive index of the medium.

The atmosphere has a refractive index which changes with height above the Earth, so that light rays seldom travel on straight paths but are slightly though continuously bent. Radio rays are similarly bent, to an extent which depends on their wavelength.

The radar horizon. If a straight line were drawn from a point at a given height above the sea, tangential to the Earth's surface, its point of contact with the Earth would be the *geometrical horizon* from the point of origin. Since light rays do not travel on straight lines but are slightly bent towards the Earth, the eye at the same height above the surface would see the *optical horizon* at a somewhat greater distance; it would see slightly 'over the curve of the Earth.' With radio rays, the bending being greater, the horizon distance will be greater than in the optical case. With a standard atmosphere, which will be defined later, the *radar horizon* for 3-cm. waves exceeds the optical horizon by about 6 per cent. and the geometric horizon by about 15 per cent. (Fig. 6 and Appendix III).

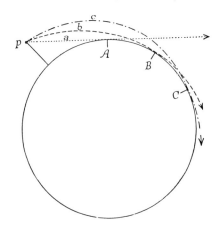

(a) The geometric tangent from p to Earth's surface, producing geometric horizon at A
(b) Optical ray from p is slightly bent, producing optical horizon at B
(c) Radar ray from p is bent more, so that radar horizon occurs at C

Fig. 6. *The geometrical, optical and radar horizons.*

This is important because the sum of the radar horizon distances of the aerial and of a target defines the maximum possible distance from which that target could return an echo. A table calculated for various heights is given in Chapter 8.

Diffraction. When a wave is partly obstructed by some object it is diffracted. The effect of diffraction is to bend the wave round behind the obstruction. The shadow cast by the obstruction will not be complete in the geometrical sense; the width of total shadow will be less, but the width of partial shadowing may be greater.

RADAR PRINCIPLES AND GENERAL CHARACTERISTICS 13

Reflection, absorption and target response. When light waves come in contact with matter they are transmitted (i.e. passed on), reflected, scattered or absorbed. Precisely the same applies to microwaves, although their detailed behaviour may be different because of their much greater wavelength. Reflection or scattering of radio waves is in fact *re-radiation*; whether the reflector is the scanner, a target or the surface of the sea, it acts as an aerial and re-radiates some of the energy which strikes it. The factor which is of the greatest importance in a radar target is the proportion of the energy received which it will send back in the direction of the scanner. This *echoing characteristic* depends mainly upon the size, shape, aspect and surface texture of the target and is examined in detail in Chapter 4. Any object which is capable of absorbing or re-radiating radio energy must to some extent reduce the power in the onward going wave; beyond the intervening object there will, therefore, be less power available to detect targets, while echoes from the targets will also lose energy on their return path.

It will be realized that no target, other than one which is very large and very close, will receive more than a minute proportion of the energy sent out from the aerial. Unless it is very specially shaped, therefore, it will return in the direction of the aerial only a minute proportion of the energy it receives. Hence the power in the echo reaching the aerial will be a very small quantity indeed.

THE RADAR SET

This description of wave character and behaviour has brought out the fact that the radar transmitter must produce short, high-power pulses of radio-frequency energy on a wavelength of a few centimetres, separated in time by certain minimum intervals; that the aerial or antenna must direct the radio wave in a narrow beam and rotate so as to scan the surface continuously; and that the receiver will have to make possible the use of echo pulses of extremely small power to operate a display or indicator, on which the ranges and bearings of the targets which returned them can be observed simultaneously. The details of the transmitter, the scanner and the receiver are considered in the next chapter. However, with the characteristics of the transmitted wave and the need for exactly timing the return of the echo freshly in mind, it is opportune here to consider some of the important features of the display.

The display

The object of the display unit, which is sometimes called the indicator, is to present to the observer a clear and accurate picture of the area around the ship, showing all objects in their correct positions in range and azimuth. Assuming that the receiver has amplified the echo pulse without destroying its essential shape, the main task of the display unit is to permit the time interval between transmitted pulse and returning echo to be measured in some simple and accurate way. The time taken by the wave travelling to and returning from a target 50 yards away is about 0·3 microseconds. To measure this is clearly beyond the capacity of mechanical devices and some kind of 'electronic stop watch' is needed.

The stop watch of the radar display is the *cathode-ray tube* (CRT) and its associated circuits. Essentially the CRT consists of a specially coated screen and an electron gun. The gun fires a very narrow stream of electrons at the screen, which fluoresces, or glows brightly, at the point of impact when the stream is strong enough. (Further details are contained in Chapter 2.) The electron stream is the 'hand' of the stop watch and the small bright area is called *the spot*. The electron stream and therefore the spot can be deflected across the screen at very high, pre-determined and constant speeds when suitable electric or magnetic forces are applied to it, while the brightness of the spot can be varied by altering the intensity of the stream.

The form of display used in marine radar is known as the *plan position indicator*, or PPI. Considering for the moment only the simplest form of display, known as Centred Relative, the spot when at rest is in the centre of the circular screen and is deflected radially, its direction of movement being made always to coincide with the direction of the rotating scanner (Fig. 7). The deflection commences at the moment a pulse is transmitted and shows as a radial line, known as the *trace* or *sweep*.

The speed of radial deflection is arranged to suit the maximum range of the display at the time. If this range should be 10 miles, the spot is required to reach the edge of the screen at the moment an echo would return from such a target. As the time interval would be that for a wave to travel 20 miles, the spot must move from centre to edge in this time (124 microseconds). As it moves at a constant speed, its distance from the centre at any instant represents the range of a target whose echo arrives at that instant. The echo pulse is used to brighten the spot momentarily and thus it leaves a *paint* or a small area of greater brightness in the correct position on the screen, from which the bearing and range of the target can be read directly. The material used for coating the screen is such that the bright paint takes an appreciable time to fade. This quality is known as *afterglow* or *persistence*.

As the echoes from all targets will be shown in this manner, if they are within the maximum range chosen and they return echo pulses of sufficient strength, the observer will see a plan view of the objects in the area around his ship.

Resolution or discrimination

The degree to which the paint of the echo resembles the shape of the target is of great importance, since upon that will very largely depend the ease of interpretation of the radar picture and the identification of targets. The electron stream which paints the picture is a somewhat thick brush, while the character of the radar beam is such that it provides information which is ill-defined in comparison with human vision. The degree of definition obtainable depends upon the ability of the radar to resolve the

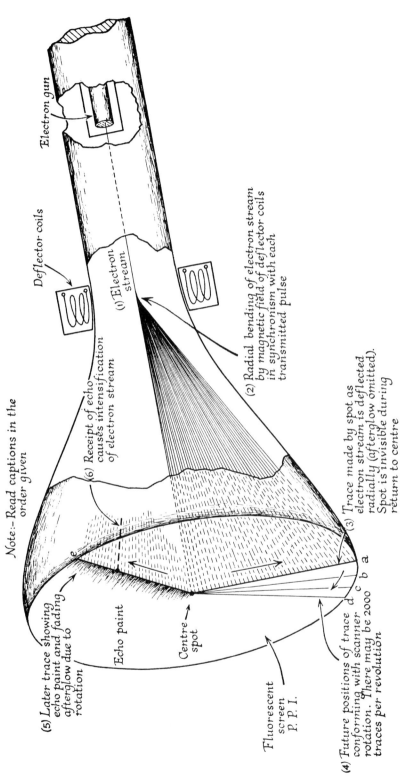

Fig. 7. The action of the PPI display.

outlines of individual targets. Apart from the detail of receiver design, the factors which limit definition are the spot size, the pulse-length and the horizontal beam-width. These can best be appreciated by examining their effects on the echo of a very small target such as a vertical post.

The *spot-size* limitation is simple: it is governed by the size of the smallest spot or the width of the thinnest line which the trace can draw upon the PPI (Fig. 8 at (*a*)); the actual area represented by the size of the spot on the PPI will depend upon the scale of the picture.

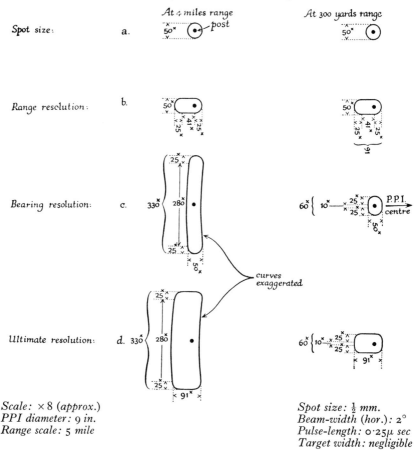

Scale: ×8 (approx.)
PPI diameter: 9 in.
Range scale: 5 mile

Spot size: ½ mm.
Beam-width (hor.): 2°
Pulse-length: 0·25μ sec
Target width: negligible

(*a*) Minimum size of a paint, due to size of spot (½ mm.)
(*b*) Minimum depth of echo (= ½-pulse-length + spot size)
(*c*) Minimum width of echo (= beam-width + spot size)
(*d*) Minimum area of resolution (combination of (*b*) and (*c*))

Fig. 8. Resolution.

If the *pulse-length* is 0·25 microseconds, its length in yards as it travels through space will be 82 yards. The post will return an echo during the full duration of the pulse so that the echo itself when returning will be 82 yards long. On the scale of the display, however, the spot moves at half the speed of the radio wave, to allow for the double journey of pulse and echo. Hence, the length of the echo of the post on the PPI will be 41 yards. This must be the minimum depth (in range) of an echo on the PPI, whatever the depth of the target (Fig. 8 at (b)).

If the *beam-width* is 2°, the post will begin to send back an echo as soon as the aerial rotation causes the fringe of the beam to strike it, that is when the centre of the beam is 1° away from the post; it will continue to return echoes of pulses while the beam sweeps across it until it is clear. Therefore the paint on the PPI will not be less than 2° in width, which, at ranges of 5, 10 and 20 miles, represents widths of 350, 700 and 1400 yards. This effect at ranges of 300 yards and 4 miles is illustrated in Fig. 8 at (c).

From this it will be clear that the radar cannot resolve an object such as a very thin post into anything resembling its actual dimensions (Fig. 8 at (d)). But it is not only with small targets that radar resolution is poor compared with that of the eye. The beam-width has the effect of drawing out the edges of the echoes of all targets in a direction at right angles to the beam, i.e. in azimuth, and so distorting them to some extent. Distortion in range also occurs but it is seldom so noticeable except at close range.

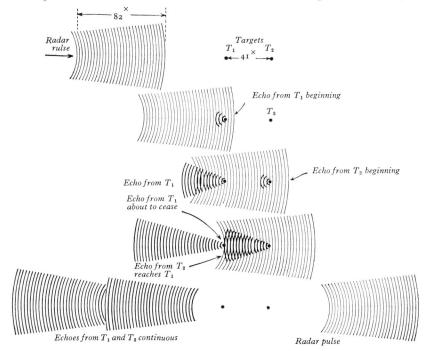

Fig. 9. Range discrimination.

The habit PPI radar has of displaying the echoes of individual targets always at something slightly greater than their proper size obviously prevents it from showing separately the echoes of objects which are close together. The slight expansion causes the echoes to merge. From the figures given above it will be seen that if two targets are on the same

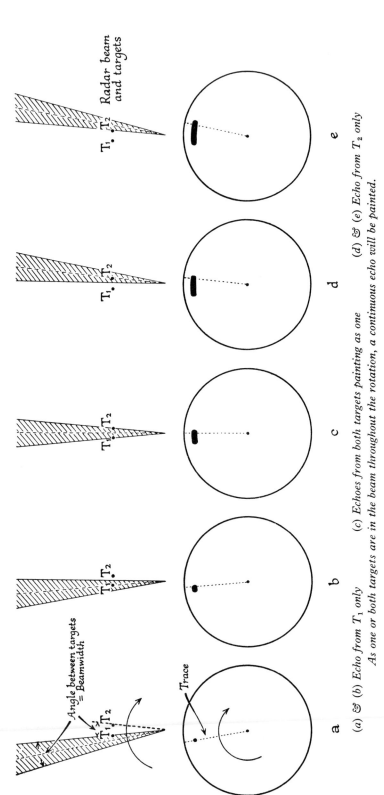

Fig. 10. Bearing discrimination.

bearing and their nearest points to the aerial are separated in range by less than half the pulse-length (41 yards in the example), the leading edge of the echo from the further target will return in time to overlap the rear edge of the echo from the nearer target and the radar will paint a single echo (Fig. 9). Similarly if two targets are at the same range and separated in bearing by less than the horizontal beam-width, overlapping will occur and a single paint be shown on the screen (Fig. 10). As the outline of whatever echo is received has to be painted with a brush of spot-size thickness, it will be clear that separation between the targets of rather more than half the pulse-length in one direction and the beam-width in the other will be necessary before the radar can show them separately. These minimum discernible target separations are called the *range and bearing discrimination* or *resolution* of the radar.

Radar equipment characteristics

Having considered radar principles in outline and examined briefly the character and behaviour of the electromagnetic waves which are employed to put these principles into effect, the relationship between the various factors mentioned and their combined effect on the picture displayed on the screen can now be discussed.

TABLE I

Characteristics of some British and U.S. 3-cm. Radars

Item	Marconi Predictor	Decca Clearscan	KH Situation Display	Sperry Mk. 16A	Raytheon TM/CA	Marconi Argus
Peak power (kW.)	25	25	25	50	50	70
Pulse length (microsecs.) (switched with range scale)	1·0	1·0	0·75	1·2	1·0	0·5
	0·3	0·25	0·25	0·25	0·5	0·07
	0·08	0·05	0·07	0·07	0·05	
Pulse repetition frequency (p.p.s.) (switched with range scale)	1000	825	800	500	900	1000
	2000	1650	1600	1000	1800	2000
		3300	3200	2000	3600	
Aerial diameter (ft.)	8	9	8	9	9	12
horizontal beam-width (deg.)	0·55	0·8	1·0	0·8	0·9	0·75
vertical beam-width (deg.)	25	15	18	20	22	20
rotation speed (r.p.m.)	24	28	24	22	33	25
PPI diameter (in.)	16	16	20	16	16	16
Range scales (n. miles)	3/8	1/4	1/2	1/2	1/4	3/4
	3/4	1/2	3/4	1½	3/4	1½
	1½	3/4	1½	3	1½	3
	3	1½	3	6	3	6
	6	3	6	12	6	12
	12	6	12	24	12	18
	24	12	24	60	24	24
	48	24	60	120	48	48
		48/60			64	
Logarithmic receiver	no	yes	no	no	no	yes
Bright display	yes	yes	yes	no	yes	no

The object is to obtain bright and clearly defined echo paints upon the radar screen; but an equipment characteristic which favours strength does not always produce good resolution. To facilitate explanation it will be assumed that a radar set employs an aerial having a horizontal beam-width of $2°$, a vertical beam-width of $30°$ and a rotation speed of 20 r.p.m.; that the pulse-length is 0·2 microseconds and the p.r.f. 600 pulses per second. The aerial will turn through $2°$ in $\frac{1}{60}$ second which is therefore the time the beam will take to pass across a point target. During this time $\frac{600}{60}$, i.e. 10, pulses will strike the target and 10 echo pulses will be returned to the aerial.

The strength of an individual echo pulse is obviously of first importance. Apart from the echoing characteristic of the target and any losses the wave may suffer in transit, this depends upon the amount of energy in the pulse which strikes the target, and hence on the power given by the transmitter and the degree of concentration effected by the aerial on the transmitted wave and on the returning echo. The choice of suitable transmitter power need not be considered here; it is governed by engineering considerations and by the maximum range required from the equipment. It is dependent on the factors to be discussed only to the extent of their combined effect on echo power. The beam-width is highly important, however. The narrower the beam the more intense will be the concentration of energy in it. If it is very narrow in the horizontal plane it will also permit good bearing resolution. If it is narrow in the vertical plane, the advantage of concentration may be lost by the target being missed due to the ship rolling. The contribution of the receiver towards the strength and resolution of the paint need not enter into the discussion at this point; neither need the rate of movement of the spot be considered.

The brightness of the paint on the PPI, however, is not dependent only on the strength of individual echoes; a building up or 'storage' effect is caused by successive echoes from the same target during the passage of the radar beam. The limiting factors of storage form a complex subject and it will be sufficient to say that the aim in radar design is usually to ensure that a target, or each part of a target, will receive 8 to 10 pulses each time the aerial sweeps past it. As was seen in the example above, the number of pulses received depends on the horizontal beam-width, the scanner rotation speed and the p.r.f. If the beam is narrowed, the rotation speed will need to be reduced or the p.r.f. increased.

The storage effect, depending as it obviously does upon accumulation of energy with time, is also governed by pulse-length. However, while a long pulse will help to build up a bright paint, a short pulse is necessary for good range resolution and a short minimum range. This conflict is often resolved by providing both short and long pulses.

Thus it will be seen that, in the ultimate display of a bright, well-resolved echo paint, all these factors are closely inter-related and often contradictory. As might be expected, the same compromise between them is not always reached. This will be seen by comparing the characteristics of typical British and United States marine equipments given in Table I.

Specialized equipments

This book is devoted mainly to radar of the conventional type but it should not be supposed that it is descriptive of all marine radar. Brief mention should be made of some specialized sets.

River radar. This may be distinguished superficially from normal radar mainly by the absence of long-range facilities. However, the requirements for short minimum range, discrimination, accuracy, absence of side lobes and other indications of refinement are even more pressing than with conventional types. The CRT used must have the best possible focusing characteristics and the best compromise between long afterglow and echo smudging.

A rate-of-turn indicator is an almost mandatory ancillary to River radar to permit the negotiation of turns in the rivers without the continual diversion of the pilot's attention by having to give helm orders.

Ship-shape radar. The need to discern the aspects of ship targets as well as the requirements in other fields such as airfield work has resulted in the development of an 8 mm. radar, of which the airfield type is in operational use. It has a beam-width of $0.15°$ and when given a trial at sea, gave a good indication of the aspect of fairly large vessels at 6 miles. Its extraordinarily high definition is seen in the photograph taken at London Airport (Plate 3). It has severe limitations in range and suffers great attenuation in precipitation.

Plotting facilities. These may be an integral part of the radar equipment or a separate ancillary. Examples of the former are the Reflection Plotter, the Photographic Radar Plot (PRP), the Automatic Relative Plot (ARP), the Autoplot (Barr & Stroud) and the TM/CA. In the latter category are the RAS Plotter and the Bial Plotter. All except Autoplot are for manual plotting. The fully automatic plotters may be in either category. They are dealt with in the Annex.

2

THE RADAR EQUIPMENT

THE object of this chapter is to give the reader a sound appreciation of how the radar set works, without requiring him to possess more than an elementary knowledge of electronics.

It has been shown that the essential units of a radar equipment are transmitter, scanner, receiver and display. These are functionally rather than physically separate units: usually, part of the receiver is contained in the transmitter box and the remainder in the display unit. They will be dealt with in this chapter without particular regard to their physical positions.

THE TRANSMITTER

It will be recalled that the function of the transmitter is to produce a succession of short, square, powerful pulses of radio-frequency energy, accurately regulated in length and occurring at a pre-determined repetition frequency. Thus, the valve which produces these radio-frequency oscillations, a *magnetron*, is required to oscillate only for very brief periods separated by much longer intervals during which it is quiescent. The transmitter circuits therefore have the duty of building up a store of energy, holding it until the moment when the pulse is due and then releasing it to the magnetron. These circuits are so designed that, after the sudden initial rise, the discharge is at a constant rate and is completed in the exact time of the desired pulse-length; that is to say it takes the form of a short square d.c. pulse. The magnetron bursts into oscillation when the energy is released and abruptly ceases to oscillate when the store is exhausted.

The components responsible for this cycle of operations, each of which is described in more detail later, are as follows. The source of energy is usually an *alternator*; the circuit which stores the energy is known as a *delay line*, a *discharge line* or a *pulse-forming network*; the holding and release of the energy is arranged by the *modulator* which is connected between the delay line and the magnetron (Fig. 11). During the charging and holding period this modulator has a very high resistance but when the moment for release arrives this resistance is broken down by the action of the *trigger circuit*. The timing and repetition of this action (i.e. pulse repetition frequency) may be governed by the alternator frequency or by the design of the trigger circuit. The action is sometimes called *firing* the modulator, an expression which aptly conveys its sudden character. The final act in the sequence is the bursting of the magnetron into r.f. oscillation on receipt of each d.c. pulse of energy. The exact methods of achieving these results may differ, but the principles described are generally applicable.

THE RADAR EQUIPMENT

A little more may be said about the nature of the various components mentioned. In the majority of modern marine radars, most valves have been replaced by solid-state devices. The principal exception is the magnetron. The advantages are a reduction in space needed and heat generated and some increase in reliability.

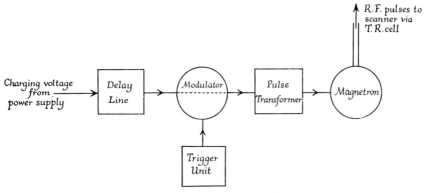

Fig. 11. *The transmitter (schematic).*

The modulator. As will have been appreciated, this device acts as a switch across which, during the discharge from the delay line, a considerable current flows. In some older high-power equipments a special low-resistance valve was used, usually a *thyratron* or a *trigatron*. In more modern sets a conventional hard valve was used. The most modern modulator is a solid-state device. In all cases the application of a pulse from the trigger circuit causes the modulator to become conductive and to allow the delay line to discharge.

The delay line is essentially a network of condensers (capacitors) and coils (inductors), the values of which govern the amount of energy that can be stored and the rate at which it will discharge. In conjunction with the modulator it therefore determines the shape of the d.c. pulse sent to the magnetron.

The trigger circuit is designed to produce a series of low-powered d.c. pulses following one another at the p.r.f. and suitably shaped for application to the modulator. The repetition frequency is governed in some equipments by this circuit itself and in others by the frequency of the alternator supplying the power.

The magnetron, in the sense that it has but two electrodes, is a simple valve. Its action, however, is complex and need not be described in detail in this book. It will be sufficient to say that its anode is a hollow cylindrical block of copper, containing a ring of cylindrical cavities in its walls. It is spoken of as a *cavity magnetron* and operates between the poles of a very strong magnet (Plates 1a and b). The cavities and the magnetic field form the basis of its unusual action. It is at present the only widely used device capable of producing high-power oscillations at microwave frequencies. When a square d.c. pulse is applied to it, it bursts rapidly into oscillation at a radio frequency which is governed by its own physical dimensions, and which is, therefore, not usually adjustable. The r.f. oscillation ceases abruptly at the end of the applied d.c. pulse (Fig. 4).

Synchronization. The basis of range measurement by radar is the time interval between transmitted pulse and returning echo. It is, therefore, necessary to set the timing arrangements in operation at the moment the r.f. pulse begins. This is usually done by using the d.c. pulse from the delay line as a trigger for the timing circuit, accepting any minute errors which may arise from inconsistency in the magnetron's action.

The aerial or antenna system

The aerial system includes the arrangements for taking the r.f. pulse from the transmitter to the aerial and the echo pulse from the aerial to the receiver. It has been explained that the power in a returning echo is extremely minute, so it will be realized that every precaution must be taken to avoid wasting the power of the transmitter pulse and the echo. The most efficient means of conveying microwave pulses from one unit to another is a waveguide.

The waveguide may be regarded simply as a pipe along which the wave may travel; it is a copper tube usually of rectangular section. The minimum cross-sectional dimensions are critically dependent upon the wavelength used and must be uniform throughout its length. It will be unnecessary to explain its action but it is important to remember that any distortion or the existence of any dirt or moisture within it will very seriously affect its performance.

Table II shows the percentage losses which occur in a 3-cm. waveguide with internal dimensions of 1×0.5 in., assuming it to be clean, dry and undistorted.

TABLE II

Length in feet	Percentage power lost in single journey
30	21 (1 db)
45	29 ($1\frac{1}{2}$ db)
90	50 (3 db)

Except in the case of the slotted waveguide aerial, the upper end of the waveguide carries a *horn* or *flare* which is suitably shaped and placed to direct the transmitted wave at the reflector and to receive the wave of the returning echo from the reflector.

Aerial design has passed through several phases in the last fifteen years. A few cheese aerials are still seen, but that design was superseded by the tilted parabolic cylinder, which was universally popular and had a better side-lobe performance. This in turn has given place to the slotted waveguide aerial, which dispenses with the reflector and horn of its predecessors and consists virtually of the last few feet of the waveguide, mounted horizontal with the narrower faces vertical. The forward looking face has a series of slots cut across it, through which the energy emerges. This aerial has a very good side-lobe performance. Some of the forms of aerial are illustrated in Plates 2a & b. The beam-width is inversely proportional to the width of the reflector or slotted waveguide which must be many wavelengths wide to radiate a narrow beam. The dimensions are critical and the aerial needs careful treatment to avoid distortion. It is usually mounted on top of a housing which contains the driving motor for rotating it and the synchronizing devices for maintaining the picture on the radar screen in correct orientation with it. The whole of this part of the equipment is commonly called the *scanner*.

The aerial may be constructed for either horizontally or vertically polarized radiation. In addition, there may be a facility for introducing circular polarization as a means of reducing rain-clutter. If fitted, this must be controlled from the display as it will tend to reduce sensitivity over the whole area. The reflection of a large proportion of the radar energy from raindrops is specular, which implies that the returned energy is reversed in phase. This does not affect the echo strength of horizontally or vertically polarized waves, but with circular polarization the direction of rotation is reversed and this will not be accepted by the aerial. This will apply not only to raindrops but to other objects which give specular reflection, e.g. radar corner reflectors.

General transmit-receive (duplexing) arrangements

The usual practice is to employ the same aerial and waveguide for transmission and reception and for the transmitter and receiver to be placed at some distance from the aerial. In such cases a considerable length of waveguide may be necessary. The loss of power in the waveguide is proportional to its length and it is desirable, therefore, to keep it as short as is compatible with obtaining a good site for the aerial. Occasionally, separate transmitting and receiving aerials are used, constructed as a single unit (see Plate 2a). In such cases the transmitter and the radio-frequency part of the receiver may be mounted immediately under it, a very short length of waveguide being run from each of them to its own aerial.

In the more usual case the waveguide is forked at its lower end, one side going to the transmitter and the other to the receiver. Remembering that the transmitter pulse is extremely powerful and the echo pulse minutely small it will be understood that the sensitive receiver must be protected from the transmitter pulse and also that the echo pulse must be prevented from wasting its energy in the transmitter portion of the guide. The protective measures may consist of placing a special valve, called a *TR cell* (or tube), in the waveguide close to the receiver (Fig. 12). Other methods of obtaining these safeguards are practised. TR cells may be tunable or *broad band*. The latter, which are capable of being effective over the whole 3-cm.

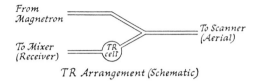

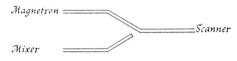

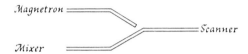

Fig. 12. Transmit-receive (duplexing) arrangement (schematic).

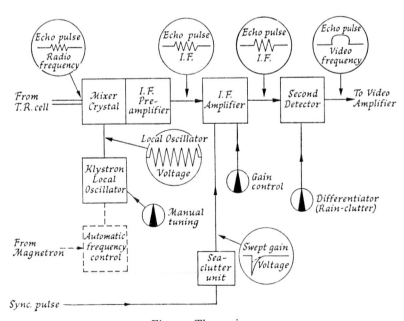

Fig. 13. The receiver.

Plate 1a. Sectional view of the cavity magnetron, showing resonant cavities.

Plate 1b. The magnetron in position, showing permanent magnet and connection to waveguide.

THE RADAR EQUIPMENT 29

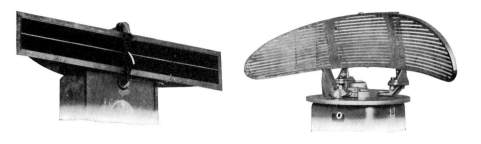

Double-cheese   *Tilted parabolic cylinder*

Plate 2a. *Two types of scanner.*

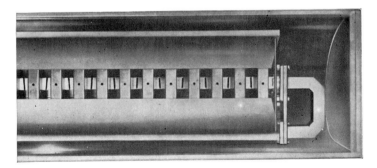

Plate 2b. *Slotted waveguide aerial, two views.*

Plate 3. *8-mm. radar at London Airport.* (*Aircraft can be seen landing on runway.*)

THE RADAR EQUIPMENT 31

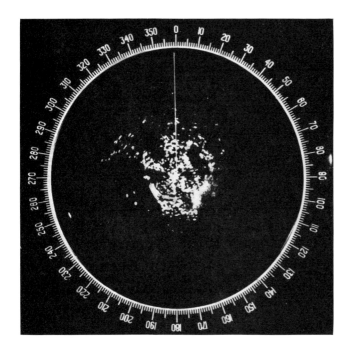

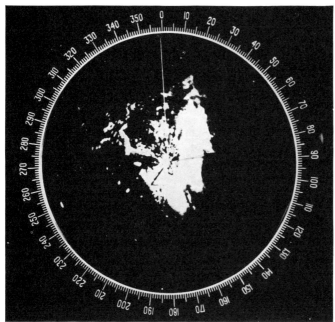

Plate 4. Use of differentiator to reduce rain clutter (at Singapore).

Plate 5. Typical display units.

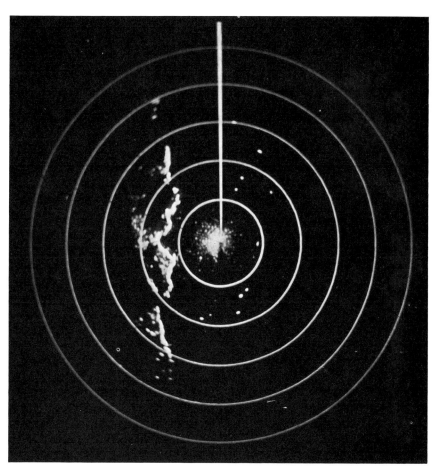

Plate 6. A PPI display showing range rings and heading marker.

radar band, avoid the necessity for re-tuning when, for example, a magnetron is replaced by one of a slightly different frequency.

THE RECEIVER

It is estimated that the power in a typical, just-detectable echo is about one-million-millionth (10^{-12}) of a watt. This is, of course, impossible to appreciate: a small torch bulb, for example, has a power of about 1 watt. A really strong echo may be a hundred million times greater than the weakest detectable one, but even so its power will still be only of the order of one ten-thousandth of a watt. The function of the receiver is to amplify, or increase, the strength of these minute echo pulses while retaining their distinctive shape, so that they will be capable of operating the display or indicator. Fig. 13 shows the arrangement of the various parts of the receiver.

No relatively inexpensive technique has yet been developed for amplifying these minute pulses at their own radio frequency (e.g. 9000 MHz). The first duty of the receiver is therefore to change the frequency of the echo pulse to one which is more manageable. When dealing with the amplification of such feeble pulses the biggest problem is probably that of keeping down the level of *noise*. The hissing noise of a sensitive radio receiver, with the volume control turned up to receive a weak signal, will be well known. In radar (PPI) presentation the noise appears as a speckled background to the picture and it must be kept to a minimum if weak echoes are to be seen.

The mixer

The part of the receiver in which the change of radio frequency is effected is called the *mixer*, of which the principal components are the *local oscillator* and the *signal crystal*. The principle is the same as that well known in superheterodyne radio receivers.

The local oscillator generates a continuous oscillation at a frequency which differs from that of the echo signal by a desired amount; this is usually of the order of 45 MHz in 3-cm. marine radar. The two are combined in the mixer by passing through the signal crystal. When two such oscillations are combined, they result in a single oscillation whose amplitude rises and falls at a frequency equal to the difference between their frequencies.

The effect of the signal crystal is to pass the echo pulse to the amplifier as an oscillation at the difference frequency (Fig. 14). This is known as the *intermediate frequency*. The signal crystal is sometimes called the *first detector*.

A few details about the components mentioned may assist understanding their operation.

The local oscillator. The valve or solid-state device usually employed is a *klystron*, which has a complex action based on the resonant cavity it contains. Since the frequency of the echo signal depends upon the magnetron and cannot be altered, any error in the intermediate frequency, due for example to replacing the magnetron by one of a slightly different frequency, can be corrected only by adjustment of the klystron frequency. A coarse adjustment may be made by a screw which alters the size of the resonant cavity, and a fine one by varying the voltage applied to one of the electrodes of the device.

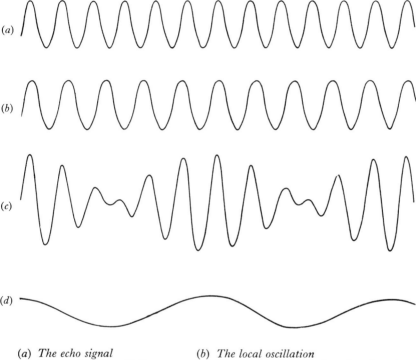

(a) The echo signal (b) The local oscillation
(c) The combined oscillation (d) The combined oscillation after detection

Fig. 14. *The principle of frequency changing.*

The frequencies of both magnetron and klystron are liable to drift slightly, particularly during the warming-up period. Slight changes in the difference frequency can be tolerated, but when they exceed a certain limit they must be corrected. Although this can be done manually, it may be more convenient to employ *automatic frequency control* (AFC) of the klystron. An AFC circuit is, therefore, often incorporated as part of the mixer.

The signal crystal is a modern version of the crystal and cat's whisker used as a detector in the early days of radio receivers. It still contains a crystal and cat's whisker but they are permanently enclosed in a sealed container.

The intermediate frequency amplifier

From the mixer the echo pulse or signal is passed to the i.f. amplifier, the duty of which is to raise its amplitude, retaining its shape and keeping noise to a low level. A radar receiver is troubled very little by interference from external sources. The *noise* already mentioned is its principal enemy and, since an echo whose strength is below noise level will not be detected, the latter governs the sensitivity of the receiver. As any noise generated in the signal crystal and the early stages of the i.f. amplifier will be amplified as much as the echo, it is obviously vital to take steps to minimize it at these points.

The i.f. amplifier usually has three controls, a *gain control*, a *sea-clutter* (or suppressor) control, and a *rain-clutter control* (differentiator), whose functions will be described.

A *logarithmic amplifier* is sometimes fitted. This is arranged so that a randomly fluctuating signal is amplified in inverse proportion to its *mean* level. Clutter gives such a signal and will, therefore, be amplified to a level which will not saturate the screen so that a stronger steady echo in the clutter area will be distinguishable. If the log amplifier is followed by a differentiator, the clutter level can be reduced substantially to the level of the receiver noise (page 122). After differentiation anti-log expansion is usually employed to restore discrimination between weak and strong signals.

Gain control. The effect of this is much the same as that of the volume control in a radio receiver. It adjusts the overall amplification so that weak echoes may be improved or eliminated as required. As will be seen later, this is of considerable importance in the interpretation of the radar picture and in the resolution of echoes.

The sea-clutter control. Waves in the vicinity of the ship return quite strong echoes (sea clutter) and, with the gain control set at the best position for detecting more distant targets, these echoes will often be strong enough to paint at maximum brightness, thus *saturating* the central area of the display and possibly obscuring the echoes of small targets within it. Returns are strongest from waves near the ship and fall off as the distance away increases. The sea-clutter control (swept gain) ameliorates this effect by reducing the gain by a decreasing amount for a short period after the pulse has been transmitted (Fig. 15). The amount of the maximum reduction and/or the time (and therefore the range) during which it is applied is determined by the setting of the sea-clutter control.

The sea-clutter circuit is one of those which are synchronized to the d.c. pulse from the delay line.

The rain-clutter control (differentiator or FTC) acts on the second detector and sharpens up and reduces the length of all echo pulses on the screen. In addition to reducing the saturating effect of rain clutter it improves discrimination generally and the definition of land echoes (Plate 4).

General arrangement. The i.f. amplifier contains a number of stages

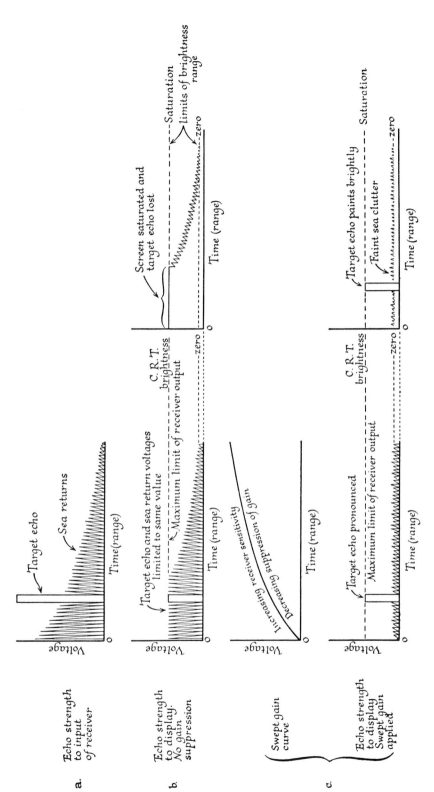

Fig. 15. *The effect of sea-clutter control (swept gain).*

of valves or solid-state devices, each one contributing to the amplification. The stages are of course tuned, though not sharply, to the intermediate frequency; a certain flatness in tuning (band-width) is necessary to preserve the shape of the pulse. Usually the first two or three i.f. stages (i.f. pre-amplifier) are accommodated in the transmitter assembly near the mixer. This is to ensure that the echo signal has a reasonable strength before it is taken to the display or indicator unit in which the remaining i.f. stages are housed. The two units may be a considerable distance apart and the echo pulses are taken from one to the other along a special cable.

To paint the echo on the screen of the PPI a momentary increase of spot brightness is required. This calls for a variation of one of the d.c. voltages applied to the CRT. For the echo pulse to provide this variation, it must be converted from i.f. to d.c. This is achieved by the normal process of rectification or detection, which is the final duty of the i.f. amplifier. A diode valve or transistor is usually employed for the purpose; it is known as the *second detector*. If the shape of the original transmitter pulse had been maintained during its travel to and from the target and in the receiving circuits, the d.c. pulse from the i.f. amplifier would closely resemble the d.c. pulse from the delay line, which began the operation. Although this ideal is not attained, it is approached closely enough for practical purposes.

The video amplifier

Some further amplification is needed after the echo pulses have been converted into d.c. pulses at the second detector. The amplifier which provides this final preparation for visual presentation is called the *video amplifier*. It has been pointed out that the powers of received echo pulses may differ by as much as a hundred million times. The maximum variation of brightness which a CRT is able to handle, without deleterious effects upon the clarity of the display, is about two to one. The variation of echo-pulse amplitude must therefore be limited to this ratio. This is partly achieved automatically by the action of the i.f. amplifier, but final limiting arrangements must be included in the video amplifier. From this unit the pulses are passed to the CRT. The successive changes which take place in the form of the echo pulse are illustrated in Fig. 13.

As will be seen later, the video amplifier is also used to amplify the pulses which are employed for painting certain of the measuring marks on the screen.

THE DISPLAY OR INDICATOR

The function of the display (Plate 5) is to show the echoes of targets on the screen of the CRT in such a way that the range and bearing of a target from the ship can be seen at a glance. It was explained in Chapter 1 that, with the centred Relative display, the PPI is a circular plan with the ship represented at the centre. It is now necessary to enlarge slightly on this description. There is another form of display, known as True motion, in which the position of own ship moves on the PPI on its correct compass course and with its correct speed to the scale in use. The technical details and operational aspects of this are dealt with in later chapters. This chapter is based on the Relative display. It is only necessary to keep in mind that own ship's position is the trace origin; if it moves, so does the trace and all the effects which depend upon it, echoes, heading marker, range rings, &c., which are explained below. On the Relative display, bearings of echo points can always be read off from an azimuth ring with the aid of a radial line, usually engraved on a rotatable cursor. Sometimes an electronic bearing cursor is provided, which is a radial line, similar to the heading marker but distinguishable by its character. The bearing is read from a separate scale. The distance of an echo paint from the centre, however, will depend not only on the range of the target but also on the scale in use; that is, on the maximum range which is displayed on the screen at any time. An automatic means of indicating range is therefore necessary.

Briefly the only way of showing anything on the screen, whether echo or measuring mark, is by brightening the spot of light at the appropriate moment during its deflection. Since the distances of both echo paints and range marks from the centre must depend on the measurement of time intervals, based on the speed of wave travel, they must both be related to the instant the transmitted pulse started. Hence the arrangements for ranging must be synchronized with the beginning of the d.c. pulse from the delay line. On the other hand the position *in azimuth* of the echo paint and any indication of bearing, such as the direction of ship's head, must be synchronized in orientation with the rotation of the scanner and sometimes also with the gyro compass. It will assist understanding of the display mechanism if these two types of synchronization are kept in mind, since most of the circuits of the display are related to one or the other.

The cathode-ray tube

The general construction of the CRT is illustrated in Fig. 16. Its main features were described in Chapter 1. The principal circuits associated with it are shown diagrammatically in Fig. 17. It will be remembered that the trace or sweep is produced by deflecting the spot radially from the centre to the edge of the tube and that it is rotated in azimuth, in synchronism with the aerial. The deflection occurs once for every transmitted pulse, that is perhaps 2000 times per second; the rotation may occupy three seconds. In this time, therefore, the spot will have passed over every part of the tube face and, if it is strong enough, the whole screen will glow brightly. To obtain a good contrast between echo paints and the background, it is desirable that the strength of the spot should be just insufficient to produce visible light in the absence of noise. When the latter is present it will then show as a faint speckled background.

The spot, of course, has to return to the centre at the end of each trace or sweep. So that it shall not paint while returning, it is brought up to the

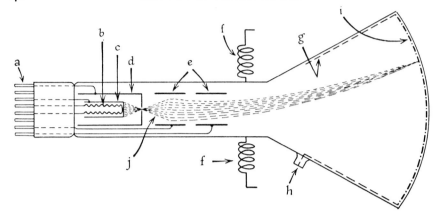

(a) Pins at base, through which connections are made
(b) Heater
(c) Cathode, emitting electrons when heated
(d) Grid, controlling number of electrons flowing along tube
(e) Cylindrical anodes forming electron lens, focusing electron stream to point on screen
(f) Deflector coils, controlling movement of spot across tube face
(g) Conductive coating on tube acts as final anode, accelerating electrons along tube
(h) High voltage connection to final anode
(i) Sensitive screen
(j) Electron stream

Fig. 16. *The construction of the cathode-ray tube.*

proper brightness during its outward travel only, by the *brightening pulse*. Its return movement, called the *fly-back*, is made at extremely high speed.

The controls and circuits associated with the display will be dealt with under three headings, corresponding with those which are unsynchronized, those which are synchronized in time (range) and those which are synchronized in azimuth. Controls which are usually mounted on the display console but which are not directly associated with the CRT's circuit are also mentioned.

Unsynchronized controls

The *gain and rain-clutter controls* are mounted on the console. They act upon the i.f. amplifier as already described. The *focus control* enables the electron stream in the CRT to be focused into an extremely narrow pencil. This ensures that the spot is as small as possible so that the picture may be sharply defined.

Time-synchronized circuits

The most important of the time-synchronized circuits is the *time base*. This circuit supplies current to the *deflector coils*, which cause radial deflection of the spot, commencing at the moment the transmitter pulse starts. According to the range scale selected by the *range switch* the time-base current rises at a rate appropriate to the speed at which the spot must

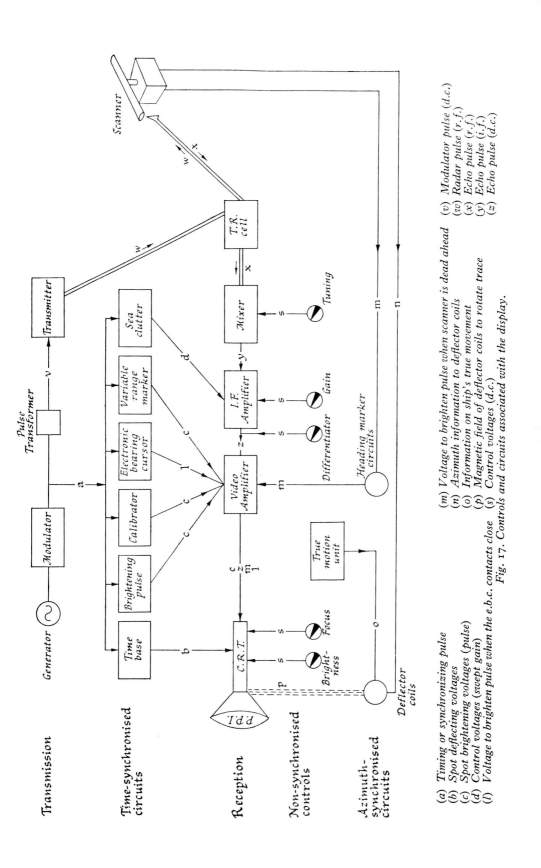

Fig. 17. *Controls and circuits associated with the display.*

(a) Timing or synchronizing pulse
(b) Spot deflecting voltages
(c) Spot brightening voltages (pulse)
(d) Control voltages (swept gain)
(l) Voltage to brighten pulse when the e.b.c. contacts close
(m) Voltage to brighten pulse when scanner is dead ahead
(n) Azimuth information to deflector coils
(o) Information on ship's true movement
(p) Magnetic field of deflector coils to rotate trace
(s) Control voltages (d.c.)
(v) Modulator pulse (d.c.)
(w) Radar pulse (r.f.)
(x) Echo pulse (r.f.)
(y) Echo pulse (i.f.)
(z) Echo pulse (d.c.)

travel. As was explained in Chapter 1, if it is required that the radius of the PPI should represent a *target range* of 20 miles, the time taken for the spot to reach the edge must be that for a wave travel of 40 miles, i.e. 246 microseconds. It is usual to have a choice of seven or more range scales to suit operational requirements; a typical selection being $\frac{1}{2}$ or $\frac{3}{4}$, $1\frac{1}{2}$, 3, 6, 12, 24 and 48 miles.

It is arranged that the spot travels at a constant speed, so that the distance along the trace from the centre at any moment will be proportional to the distance the wave has then travelled.

The brightening pulse circuit ensures that the spot is brightened during its outward travel only. The *brilliance control* governs its brightness in the absence of noise, echo pulses or range or heading-marker pulses. The arrival of any of these during the period of the trace causes an additional brightness.

The sea-clutter control, placed on the display console, operates into the i.f. amplifier. Being synchronized with the delay line d.c. pulse it commences to operate at the beginning of the trace, reducing the gain of the amplifier by a decreasing amount and for a time (range) selected by the control, and hence affects the brightness of echo paints within the selected distance from the trace origin.

Range or calibration rings. One method of measuring the range of echoes is to show bright rings on the PPI at known distances from the centre (Plate 6). The range of an echo paint which falls between the rings is measured by interpolation. One of the functions of the range switch is to select a distance between the rings convenient to the scale in use. They are produced by the *calibrator circuit* in which an oscillation is generated and shaped to give a series of sharp pulses, separated in time by an amount corresponding to the selected distance between the rings. The oscillation is, of course, synchronized with the delay line d.c. pulse. The calibrator pulses are usually passed through the video amplifier to the CRT and cause the spot to brighten at the required intervals and so to produce the rings. The rings are usually introduced at will by a switch.

The variable range marker. Often an additional circuit is included, which produces on the PPI a single ring, the radius of which can be varied by the *range marker control*. The latter also drives a drum or counter from which the range represented by the position of the ring can be read. This is obviously a more handy method of *reading* the range than interpolating between range rings, though it is necessary to calibrate it against the range rings from time to time. It is much more versatile in its operational uses. (See also Interscan markers, page 222.)

Azimuth synchronized circuits

As the trace of the PPI rotates, it displays echoes from the objects which are being 'viewed' by the aerial. To be able to determine the direction of these echoes it is clearly essential that the direction of the aerial at any moment should be known. That is to say, not only must the trace rotate in synchronism with the aerial, but also the orientation of the picture on the screen must be in correct relation with the fixed bearing ring of the PPI.

The heading marker. The bearing ring is graduated in degrees from one to 359°, o being shown at the top. The scale may represent bearings relative to the ship's heading or to a point of the compass, e.g. north: the picture has to be correctly orientated for whichever method is in use. Each time the scanner passes through the direction of the ship's head a contact in the scanner housing operates a circuit which sends a pulse through the video amplifier to the CRT. This pulse is long enough to brighten the whole of a trace, with the result that a radial line appears on the PPI (Plate 6). Since it is originated mechanically by the scanner in one position only, the position of this *heading marker* will be correct in relation to the position of echo paints. If the picture is rotated by some means until the heading marker is at 0°, the bearings of all echoes may be read off directly as relative to ship's heading. Similarly if the heading marker is placed at the point on the bearing scale representing the ship's true course, the direction of echoes may be read off directly as true bearings.

Trace-rotation mechanism. There are various methods of rotating the trace in synchronism with the aerial. In the electromagnetic CRT (the usual type in radar equipments) the spot is deflected radially by increasing the strength of the magnetic field set up by the time base current flowing in the deflector coils. The direction in which it is deflected is governed by the direction of the magnetic field; if the field is rotated, the direction of deflection of the trace will follow it round. The field is in fact driven round by a device which is kept automatically in synchronism with the aerial. This is achieved either by mechanically rotating the deflector coils or by using fixed coils supplied with current in such a manner as to cause the field to rotate. The latter is more costly and so is usually found only in the more expensive equipment. The field and hence the picture can also be rotated by hand, using the *picture-rotate control*, so that it may be correctly orientated as already mentioned.

It will be noted that if the picture were merely to be aligned by hand north upwards, it would be necessary to re-align for every alteration of the ship's course. When the ship has a gyro or any other type of transmitting compass the picture may be stabilized in orientation with the compass, by arranging that the trace-rotation mechanism is controlled by both the scanner and the compass through a differential drive.

Compass stabilization has several advantages over the fixed, ship's head upward display. However, many mariners prefer the latter. To permit the advantages of both to be realized, there are radars in which the picture is stabilized on the CRT, which itself can rotate and is stabilized separately so that the heading marker is held at 0°. (See page 173.)

Whatever form of presentation is used, it is very important that the orientation of the picture should be checked from time to time by observing the position of the heading marker. Without stabilization, the position of the heading marker has no relation to the compass; whenever a bearing is taken, ship's head by compass must be noted so that bearings may be corrected for yaw.

Electronic bearing cursor. This is far more accurate than the mechanical cursor as it removes parallax errors and those due to the trace origin not being at the PPI centre. It is a radial line on the PPI rotated by a control knob and read off on a separate scale. Being synchronized with the gyro or transmitting compass, it will register compass bearings and so eliminate yaw error. (See also Interscan markers, page 222.)

The performance monitor

An indication of the level of performance of the whole radar equipment is an important adjunct to the display. It has become common practice to use for this purpose a device known as an *echo box*. This consists of a cavity, which is resonant at frequencies in the marine radar band and which, when excited by a pulse radiated by the radar transmitter, produces a long 'ringing' echo which may be compared with that given out by a sharply struck bell. The harder the bell is struck (transmitter power), the longer will it ring; the better the hearing of the listener (receiver sensitivity), the longer will the slowly weakening ringing be heard.

The echo box is usually placed so that, during a small arc of every revolution of the scanner, it will receive pulses from the transmitter, and so that the ringing echo will be sent back into the scanner and passed down the waveguide into the receiver. The echo thus follows the same path as an echo pulse from a target and is displayed on the PPI on a bearing corresponding to the direction of the scanner at the time. As the echo will be received very shortly after the transmitter pulse finishes, it will appear from the centre of the screen and, being a long ringing echo, will extend some distance outwards. This gives it the appearance of a narrow plume and it usually extends to a range of 1000–2000 yards. The length depends upon several factors including transmitter power and receiver tuning and sensitivity and it is thus a measure of the overall performance of the equipment. If its length is noted when the set is known to be working at full efficiency, any deterioration can readily be detected. (See Chapter 3.)

The echo box is usually placed so that the echo occurs inside a blind arc, e.g. in line with the funnel or, failing this, on a bearing on which the detection of target echoes is relatively unimportant, e.g. astern.

Other systems are used. In one, the transmitter monitor receives energy from a neon lamp mounted in a horn across which the radar beam sweeps. A push button at the display applies an ionizing voltage to the neon lamp, in which the amplitude of the current is proportional to the power radiated from the aerial as it sweeps across it. This energy is fed into the video receiver and appears on the display as a plume.

The receiver monitor obtains its energy from an echo box coupled to the waveguide near the magnetron. This is fed to the receiver input and appears on the display as an all round response, the radius of which is proportional to receiver sensitivity.

THE COMPLETE EQUIPMENT

Fig. 17 illustrates the arrangement of the parts of a complete equipment. This and the brief description given should be sufficient to enable the reader to follow logically the sequence of operations and to permit full appreciation of the chapters on controls and interpretation.

3

OPERATIONAL CONTROLS

CONTROLS are of two kinds; those which are immediately accessible and those which are not. Controls of the latter kind are those which seldom require adjustment after the initial setting-up, for which reason they are usually called pre-set controls. The best results cannot be obtained from any radar unless all the controls are in proper adjustment.

Incorrect operation of the controls may lead to failure to detect ships or buoys, or to inaccurate measurement of range or bearing. It is therefore very important that the user should understand the function of all the controls, know where they are, how to adjust them correctly and how to find out whether they are in correct adjustment. In many cases the correct adjustments vary from time to time. For example, the settings of the BRILLIANCE and FOCUS controls inevitably change as the cathode-ray tube deteriorates with age, and the setting of the GAIN control should always be re-checked after adjusting BRILLIANCE or changing range. In some sets the centre of the display moves slightly every time an alteration of course is made. Never to make any readjustments is just as dangerous as the unnecessary alteration of pre-set controls. It is advisable to institute a routine check of the overall performance, of the centring of the display, of the direction of the heading marker and of the accuracy of the range-measuring facilities at the beginning of every spell of watch-keeping, and to practise during good visibility the adjustments of controls which may be of vital importance in fog.

In addition to the on-off switch, the controls and facilities which may be required during the operation of a marine radar include:

For the adjustment of performance—
- BRILLIANCE
- FOCUS
- GAIN
- SEA CLUTTER (SUPPRESSOR)
- RAIN CLUTTER (DIFFERENTIATOR)
- TUNING

For checking the performance—
- PERFORMANCE MONITOR

For picture alignment and for the measurement of bearing—
- BEARING CURSOR (MECHANICAL AND ELECTRONIC)
- CENTRING
- HEADING MARKER

OPERATIONAL CONTROLS

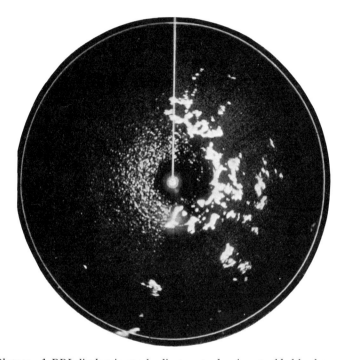

Plate 7. A PPI display in good adjustment, showing speckled background.

PICTURE ROTATE
EXPAND CENTRE
SCALE ILLUMINATION

For measurement of range—
RANGE SCALE
RANGE RINGS
VARIABLE RANGE MARKER

For Azimuth Stabilized Displays—
PRESENTATION SWITCH e.g. Ship's Head Up or North Up

For True Motion Displays—
PRESENTATION SWITCH e.g. Relative Motion or True Motion
SPEED INPUT SELECTOR
SPEED ADJUST
COURSE ADJUST
N–S AND E–W RE-SETTING
SET AND DRIFT CORRECTION

The names given to these controls may differ somewhat between one set and another, and on some radars the names are replaced or supplemented by internationally recognized symbols. Appendix VIII shows the symbols in current use.

Once the initial setting-up has been made, it is likely that during an ordinary spell of watch-keeping only a few on this long list of controls will need adjustment. Some of the controls are normally used only in the setting-up process, and are made accessible to facilitate compensation for changes in components or in external conditions. Those used most frequently are the BEARING CURSORS, RANGE SCALE, RANGE RINGS and VARIABLE RANGE MARKER. The PERFORMANCE MONITOR, of course, will be under observation from time to time.

Adjustment of the display

The brightness of the radar display is controlled not only by the BRILLIANCE control, but also by the receiver GAIN, and both must be used when making a working adjustment of the level of brightness.

The definition or sharpness of the display is adjusted by means of the FOCUS control. Preliminary adjustments of brightness and focus are made with the receiver GAIN at minimum. The BRILLIANCE control is set so that the rotating trace is clearly visible, but not too bright. The FOCUS control is then adjusted so that the trace is as sharply defined as possible.

The BRILLIANCE control is then turned until the rotating trace is just invisible, and the GAIN increased until a speckled background appears on the screen. This represents the condition of optimum sensitivity which should be used in normal circumstances. It is more easily recognized on the longer range scales than on the shortest (Plate 7). Unless the speckled background is visible, weak echoes will not be detected. On the other hand, if the brightness level is so high that the afterglow is not speckled but uniformly bright, the screen is 'saturated,' and no echoes at all will be detected. The life of the cathode-ray tube decreases as the average brightness is increased; its level should, therefore, be the minimum appropriate to the range scale in use, the ambient light and the observer's visual comfort.

When the range scale is switched from one scale to another, the brightness level usually changes and must be re-set. After the initial setting-up process, any necessary change can usually be made with the GAIN control alone. For this reason, the BRILLIANCE control in some equipments is pre-set. It will, however, require adjustment from time to time during the life of the cathode-ray tube, or if there is any large change in the level of ambient light.

After the gain and brilliance controls have been set to the optimum working level, a final adjustment of the FOCUS control should be made. The range rings are good subjects for focusing. One of the longer range scales should be used during this adjustment. The spot is liable to defocusing to an extent which varies with its distance from the centre of the tube. A ring about half-way between the centre and the edge of the tube is therefore normally used to ensure a reasonable focus everywhere on the display.

Gain and sea clutter

The SEA-CLUTTER control should normally be turned down when the sea is calm. In a rough sea it will almost certainly be found that, even with the most careful adjustment of the control, there are small zones of saturation, and also zones which are dark, showing no clutter echoes or afterglow speckling. The dark patches are regions in which the receiver gain is too low. A common fault in setting this control is to have too much suppression, so that there is a zone of darkness around and beyond the maximum range to which clutter extends. This may be dangerous, particularly if the GAIN is set so that the speckled afterglow background is not clearly visible at longer ranges, because it means that a target at or just beyond the limit of the sea-clutter range may not be detected. It is preferable for the adjustment to be correct at the outer part of the clutter zone, some saturation at the closer ranges being accepted. In any event, small variations of gain will still be necessary after the sea-clutter control has been set; the gain must be decreased periodically to examine the zones of saturation, and increased from time to time to examine any dark patches.

Rain clutter

Echoes from a heavy rainstorm also can saturate the display; it is invariably found, however, that such rain falls over areas of limited extent, which are seen on the radar display as discrete patches of rain clutter. In examining an area saturated by rain clutter which may be obscuring the echo from a ship, gain is periodically reduced until the rain clutter is just visible for long enough to enable any ship echo to be detected. (Also page 122.)

In most radar equipments a circuit is included which automatically reduces saturation of the screen by rain clutter. This is controlled from the display by a switch or control marked DIFFERENTIATOR or FTC or RAIN-CLUTTER. Effectively this circuit breaks up very large blocks of echoes such as are produced by rain clouds. It will also do this for any area of saturation including large land masses, and is particularly useful when using the radar for short-range river or harbour navigation.

Tuning

In some marine radars the receiver is automatically tuned to the transmitter by means of the automatic frequency control circuit. In most, however, manual tuning only is provided; these sets have tuning meters to indicate when the receiver is in tune. Where AFC is included, manual tuning is sometimes provided as an alternative which may be switched in if the AFC circuit becomes defective.

Without AFC the tuning of a radar receiver tends to drift for ten to twenty minutes after the set is first switched on. During this period it is necessary to re-tune manually at intervals of a few minutes, but thereafter no further adjustment is normally required. It is important, however, to check the tuning from time to time, particularly if there are no echoes visible on the screen.

The echo-box PERFORMANCE MONITOR is a most valuable accessory, since it not only provides an echo which can be used for tuning the receiver but also enables overall performance (including tuning) to be checked merely by inspection of the display.

Tuning at sea is often possible even when there are no ships or land targets in the vicinity, no performance monitor and no tuning indicator. It may be done by turning down the anti-clutter control and tuning for maximum density and extent of sea clutter. Even when the sea is calm, echoes from the surface usually extend to a range of a few hundred yards. During this operation a range scale appropriate to the extent of the sea-clutter area should be used. This procedure can be carried out on any set on which the minimum range is short (e.g. 50 yards).

True motion

True motion presentation is usually an alternative to relative and may be switched in and out as desired. For True motion the display requires to be stabilized from the gyro compass or transmitting magnetic compass. Own ship, of course, remains identified with the point from which the trace originates (cf. centre spot on relative displays) and round which the range rings and variable range marker are formed and from which the electronic bearing cursor stems.

OPERATIONAL CONTROLS

There are controls for setting the ship's course and speed; the speed may be set manually to the estimated speed of the ship through the water or it may be fed in from the ship's log. There is a control for direction and speed of current, unless this is done by a 'course made good' control and there are re-set controls for adjusting position of own ship on the PPI. Because own ship moves across the PPI, it is necessary from time to time to re-set own ship on the PPI to avoid loss of view ahead. On most True-motion radars this will occur automatically when own ship reaches two-thirds of the cathode-ray tube's radius. Where re-set is not automatic, an audible or visual warning is given to prevent it moving too far and so losing the view ahead. There may be a control to re-set this warning.

In the true-motion display, as in others, there is the possibility of making mistakes, and it is desirable to understand exactly what the movement of own ship and echoes on the screen may mean in various conditions of adjustment. There are three main conditions which can be introduced by adjustment of the true-motion controls; for purposes of explanation these might be called 'ground-stationary', 'sea-stationary' and 'ship-stationary'.

If own ship has way on, the ship-stationary condition on the PPI is reached by putting the true-motion speed setting to zero. All echoes will then behave exactly as they do in a relative display. This condition may be useful when relative track is required.

The sea-stationary condition is reached when own ship's heading and speed through the water are set on the controls. In this condition the echoes of land, fixed marks, anchored ships, &c., will move in the opposite direction to any set or drift which the ship may be experiencing and will show appropriate echo trails. The echoes of other vessels under way will move according to their *heading and speed through the water*; this assumes that all vessels in question are experiencing the same set and drift. This therefore will be the best condition in which to operate in open water.

Finally there is the ground-stationary condition. True-motion displays provide means of allowing for set and drift (together), and when the adjustment has been made correctly the echoes of land and other fixed objects will be stationary. In this condition the movement of own ship on the display and the echo trails of other moving objects will represent *course and speed made good*. This condition may be preferred when moving in rivers or narrow estuaries. In these circumstances, however, the amount of cross-set is likely to be variable and not large; the most practical solution may be to adjust the speed setting to allow for current with or against the ship and to neglect cross-set. This of course will mean that there may be some slight residual movement of land echoes from time to time, but this is not likely to be serious.

It will be seen that the only difference between the ground-stationary and sea-stationary conditions is the matter of the current. It may be important to remember that an estimate of another vessel's speed in a river with a strong current will be affected very much by the latter in the ground-stationary condition. Thus if two vessels on a river are moving at 10 knots (through the water) in opposite directions and both are experiencing a 5-knot current along the line of advance, the rate of progress of the two ships as measured by their echo movement will be 5 knots and 15 knots respectively.

Performance monitor

The echo-box monitor was described in the previous chapter. As the length of the plume depends on the overall performance of the radar, it will not be at its maximum unless the receiver is in tune with the transmitter. Checks of overall performance should be made with gain at maximum and the anti-clutter control in the OFF position; if the latter is on and causing receiver gain to be low over the range of the performance monitor, the length of the plume will be reduced accordingly. This effect may therefore be used to check the performance of the sea-clutter control over a limited range as it is varied.

If the echo box is so sited that its plume is not in a blind arc, it may appear against a background of sea clutter. If sea clutter extends out to or beyond the length of the plume, the end of the plume will be difficult to distinguish. In rough seas, therefore, a monitor so sited may not be a reliable guide to overall performance. An apparent deterioration should be confirmed when the sea has become calm again.

Measurement of bearing

For measurement of bearing a mechanical bearing cursor is usually fitted; when this cursor is rotated, its outer edge moves over an illuminated bearing scale. It is made of thick transparent plastic, and is mounted well off the face of the cathode-ray tube; to avoid serious reading errors due to parallax, radial index lines are scribed on both sides of the cursor. The viewing position when a bearing is being measured should be such that these lines appear as a single line bisecting the echo.

If the bearing measurement is to be accurate, the centre of rotation of the trace must lie on the axis of rotation of the bearing cursor. The centre may change with changes in valves, &c. (very much less with transistors), and is always liable to move when an alteration of course is made, because of the effect of the Earth's magnetic field on the cathode-ray tube beam. In some types of radar the changes are very small, but in others they are often large. The accuracy of centring should be checked frequently. The effect of inaccurate centring is an error in bearing which becomes more serious as the echo gets nearer to the centre of the screen. This is illustrated in Fig. 18 in which *a, b, c, d, e* represent the positions on a display of echoes from a ship approaching from right ahead. The centre of the display O is displaced athwartships from the centre of the bearing cursor C. It will be seen that the echo will appear to open in bearing while the range is closing, although the bearing of the target is in fact right ahead and steady.

In practice the displacement due to alteration of course is unlikely to exceed 3 mm. This will produce a bearing error of less than 1° at the edges of the screen, rising to 9° at a distance from the centre of one-fifth of the radius of the screen. The error will be at its maximum when the bearing of the echo is at right angles to the direction of the displacement.

Horizontal and vertical CENTRING (shift) controls are used for centring the display, and are accessible to the operator in some types of set. Because of the limited accuracy with which centring can be achieved, there will always be a liability to bearing errors near the centre of the display. When accurate bearings are required a range scale should therefore be chosen which places the echo as far as possible from the centre of the display.

The electronic bearing cursor, described in Chapter 2, is a far more accurate device. It is a necessity when using True-motion since the display is not centred. In some radars a fixed bearing graticule is fitted.

To orientate the display correctly for the type of presentation desired, it is usually necessary to set the heading marker by means of the PICTURE-ROTATE CONTROL every time the set is switched on unless the radar incorporates an automatic alignment facility. If the ship has a transmitting compass the display can be stabilized in orientation by turning the MANUAL/COMPASS control to COMPASS. The alignment of the picture should be checked every time a bearing is taken. The heading marker is generally provided with a switch; it should normally be ON as a permanent monitor of the orientation, and switched OFF only for very short periods when necessary to check that the marker is not obscuring an echo.

A control of the brightness of the heading marker is not usually provided, although in some circumstances it would be an advantage.

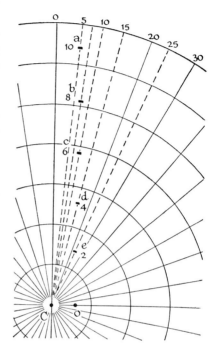

Fig. 18. Bearing errors caused by inaccurate centring.

Heading marker check procedure

In connection with bearing accuracy, experience has demonstrated the need to exercise particular care in checking heading marker alignment.

Visually aligning the aerial along what appears to be the centreline of the ship is not a sufficiently accurate method. The following procedure is recommended:

(i) Adjust accurately the centre of rotation of the trace; switch off azimuth stabilization.
(ii) On equipment having appropriate controls, rotate the picture so that the heading marker lies at zero degrees on the bearing scale.
(iii) Select a conspicuous visible object whose echo is small and distinct and lies as nearly as possible at the maximum range of the scale in use. Measure simultaneously the relative visual bearing of this object and the angle on the PPI which its echo makes with the heading marker; the visual bearing should be taken from a position near the radar aerial. Repeat these measurements twice at least and calculate the mean difference between bearings obtained visually and by radar.
(iv) If an error exists, adjust the heading marker contacts in the scanner assembly to correct the position of the heading marker by moving it an amount equal to the mean difference calculated in (iii) above.
(v) Rotate the PPI picture to return the heading marker to zero on the bearing scale.
(vi) Take simultaneous visual and radar bearings as in (iii) above to check the accuracy of the alignment. The procedure should employ positively identified targets whose echoes are clear of confusion on the PPI, otherwise serious bearing errors may be introduced.

Measurement of range

The size of the field of view displayed on the cathode-ray tube is selected by means of the range-scale switch. For range measurement, range rings and sometimes also a variable range marker are provided. In the latter event, the range rings are provided for calibration of the variable marker, but are not always usable for range measurement.

When using the variable marker, the usual convention is to align it so that the outer edge of the marker just touches the inner edge of the echo. If the variable marker has been calibrated by means of the range rings an accuracy of about $2\frac{1}{2}$ per cent. of the maximum range of the scale in use can be obtained; unless this calibration is done at the time of measurement, the accuracy should not be assumed to be better than 5 per cent. Measurements at ranges of less than about 300 yards are liable to errors of 50 yards or more, and the range accuracy of an equipment at the shortest ranges should be determined by visual observation.

SUMMARY OF ADJUSTMENT PROCEDURE

Initial setting-up

(1) Switch on the radar in accordance with the manufacturer's instructions.

OPERATIONAL CONTROLS

(2) Set GAIN to minimum, SEA CLUTTER to OFF, RANGE SCALE to long or medium.
(3) Set BRILLIANCE until trace is visible, but not too bright.
(4) Adjust FOCUS.
(5) Set BRILLIANCE so that trace is just not visible.
(6) Set GAIN so that speckled background is visible, but not too bright.
(7) Switch RANGE SCALE to a short range; if tuning is automatic, check tuning by PERFORMANCE MONITOR (if fitted). If tuning is manual, adjust TUNING CONTROL by tuning meter (if fitted) and finally for maximum length of indicator of performance monitor. Compare this with the recorded maximum. If no indicators are provided, tune for maximum extent of sea-clutter echoes.
(8) Adjust SEA-CLUTTER control until the areas of saturation (bright patches) and of inadequate gain (dark patches) are each eliminated as far as possible.
(9) Switch RANGE SCALE to required range, check FOCUS, and adjust brightness (if necessary) by means of the GAIN control.
(10) If True motion is available, switch to TRUE or RELATIVE as desired.

In Relative motion:
(11) Check centring, and adjust if necessary by means of CENTRING controls (if fitted).
(12) Switch on HEADING MARKER: if heading upward unstabilized presentation is required, set heading marker to zero of bearing scale by means of PICTURE-ROTATE control. Alignment is sometimes automatic.
(13) If north-stabilized picture is required, set HEADING MARKER to direction of ship's head, and switch MANUAL/COMPASS control to COMPASS.

In True motion:
(14) Select method of speed input desired and adjust speed setting as appropriate.
(15) Set ship's course.
(16) Select desired starting position for own ship by N–S, E–W controls.
(17) Adjust 'set and drift' correction as necessary (these may be marked variously, Tide, Current, &c.).

Note: When the radar is switched off, controls should be left in normal working positions, and on one of the shorter range-scales (e.g. 3 mile), so that a useful picture may be obtained in emergency immediately it is warmed up.

Routine check procedure
Range scale, gain setting, sea clutter, tuning, centring, heading and overall performance should be checked at the beginning of each spell of watchkeeping. Unless a performance monitor is fitted, overall performance cannot be checked, and tuning must be checked on sea clutter or other echoes, or by reference to the tuning meter.

4

PROPAGATION OF WAVES AND RESPONSE OF TARGETS

PREVIOUS chapters have given an outline description of what happens inside a radar set, traced the action of forming the radar pulse which is fed to the aerial and followed the echo pulse from the moment of its reception at the aerial to its eventual display on the PPI. This chapter deals with the travel or propagation of the radar pulse through space, the echoing response of the objects which it may encounter and the effect of this on the strength and behaviour of the echo received back at the aerial.

The reader may find this chapter rather difficult going as it introduces a number of unfamiliar conceptions which are not easily grasped, but an understanding of them will be of the greatest value to the mariner in helping him to appreciate the reasons for the performance which his equipment will give under various conditions. This subject is usually inadequately treated and it is perhaps because of this that misconceptions are so often voiced by otherwise well-informed users of radar.

The reason why a prominent lighthouse may give only a poor echo is often a source of mystery. The fact that the stronger of two echoes may be the first to be lost as the ship steams away is not always appreciated and is seldom understood. This chapter sets out to give an understanding of effects such as these and to explain some of the general principles of echo behaviour.

An understanding of the basis on which the chapter has been written may help the reader to follow it. The overall picture of echo behaviour is a complex one built up from a number of fairly involved considerations. Wherever possible, each new part of this chapter starts with a simple assumption which may not represent exactly the actual conditions met in practice. This simplified assumption is then developed, and, one by one, the various complications caused by the actual conditions are introduced, thus making the approach easier than if the actual conditions were considered from the start. The chapter is divided into two distinct parts so that the reader may first appreciate how and in what strength the radar pulse may reach a target and return and then examine all the factors which make for good or bad targets. The chapter is written expressly for the enthusiastic *user* who wishes to understand the performance of his equipment. The engineer or physicist may find some points of use to him in it, but he must not expect it to be written in his language.

I—WAVE PROPAGATION

PROPAGATION OF RADIO WAVES

From our point of view we are concerned with the way in which the propagation of a radio wave from a transmitter to a distant point governs the strength of signal at that point and how this and the propagation of the returning echo from a target governs the strength of echo. We are concerned with how this strength is influenced by change in target position, in range, bearing and elevation.

Decrease of transmitter signal strength with range

When a transmitter is feeding a non-directional aerial, radio waves will spread out from this aerial equally in all directions as depicted in Fig. 19. Since the same signal, as it moves outward, is spread over an ever increasing circumference, it is clear that the strength of the signal as measured by an observer at any point will decrease as the point moves away from the transmitter. It is usual to refer to the strength of the signal at a point as the field strength and it is usually measured in volts (or fractions of a volt) per unit of length.

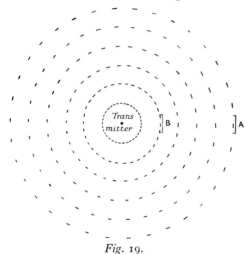

Fig. 19.

The observer at A in Fig. 19 is at 3 times the distance from the transmitter of the observer at B; he is also on a circle having 3 times the circumference of that at B and hence will experience a third of the field strength experienced by observer B. From this it may be seen that the field strength in *volts* is inversely proportional to the distance from the transmitter.

The foregoing has considered the signal spreading out in one plane only, say the horizontal plane; in fact of course it would spread out also in the vertical plane. In this case we should think of the circles as spheres and consider the signal as being spread out over the *area* of the surface of the sphere. The strength is then usually referred to as the *power* of the signal and it is usually measured in watts (or fractions of a watt) per unit of area. The observer at A in Fig. 19 is, then, on the surface of a sphere having 9 times the surface area of the sphere of observer B and will hence experience a ninth of the power experienced by observer B. From this it may be seen that the power of a signal in watts is inversely proportional to $(\text{distance})^2$.

In practice it does not much matter whether we refer to the strength of a signal in terms of its field strength in volts or its power in watts as long as we remember that we are using different types of units. In much the same way, it doesn't matter whether we compare the size of two spheres by saying that one has 3 times the diameter of the other or by saying that one has 9 times the surface area of the other. Clearly it does not mean anything to say that one sphere is '3 times the size' of the other unless we say whether we are talking about its diameter in feet or its surface area in square feet; we can say loosely that one sphere is bigger than another but we cannot say how much bigger unless we specify the type of unit. Similarly in the radio case we may say loosely that one signal is stronger than another but it doesn't mean anything to say that it is '3 times as strong' unless we say whether we mean the field strength in volts or the power in watts; if we mean that the field strength in volts is 3 times greater, then we could equally say that the power in watts is 9 times greater. Power in watts is proportional (not equal) to the square of field strength in volts.

We may summarize what we have said about the way the strength of a signal received at a point varies as we change the distance of the point from a transmitter as follows:

Field strength in volts is inversely proportional to distance.
Power in watts is inversely proportional to $(distance)^2$.

The same relationships hold even if the signal is sent out in a sharp beam so long as the width or area over which we observe the strength is small compared with the width or area of the beam at that position.

Attenuation and shadowing

It should be made clear that we have not been talking about any loss, or *attenuation*, of signal but only about its redistribution. If it were possible to have a transmitter and receiver working under theoretically ideal conditions, that is well away from the Earth and its atmosphere, out into a completely loss-free space, we would find that the above relations were exactly true. Under practical conditions if there is attenuation, the signal strength will decrease more rapidly with increase of distance than under ideal conditions. Fortunately the attenuation due to the atmosphere is normally quite small for radio waves of a wavelength of 3 cm. or greater; the loss is in fact so small within the distances covered by marine radar that we may completely disregard it for our purpose.

We may thus say that provided radio signals do not impinge on any solid (or liquid) objects to affect them, their strength will decrease with range according to the relations given for free-space conditions. If the path of the signals, however, is obstructed by, say, rain, fog or a hill there will be an appreciable attenuation of the signals beyond the obstruction. We could say that the obstruction causes a radio shadow behind it, the intensity of the shadow depending on the nature of the object causing it, in the same way that smoke issuing from a funnel will cast a slight shadow in the sunlight and the funnel itself will cast a very intense shadow. The effects of

PROPAGATION OF WAVES AND RESPONSE OF TARGETS 59

the attenuation caused by various meteorological factors are dealt with in Chapter 5. In the case of most large objects (ships, buildings, hills) the attenuation is virtually complete, that is they will cast a complete radio shadow behind them in which the signal strength will be almost zero.

Decrease of echo strength with range

Returning to free-space conditions, let us examine how the strength of an *echo* received back at a radar set will vary as the distance of the *target* from the radar is changed. Consider first the field strength of the signal impinging on the target as it is moved away from the radar transmitter. This will decrease as before; that is to say if the target moves 3 times as far away it will experience one-third of the previous field strength. Assume that it re-radiates all of the field impinging upon it, then we can regard it as a little 'transmitter' having one-third of the strength (voltage units) at the more distant position of that which it had at the closer. Now this little transmitter is sending out signals which spread out as they progress and the fraction of the transmitted signal which is picked up by the radar receiver will decrease with range exactly as in previous examples; hence when the target is moved to the more distant position the radar receiver will experience only a third of the field strength which it would have experienced for the nearer position even if the strength of the target radiation had remained constant—in fact the target is re-radiating a signal of only one-third of the field strength when it is in this position and hence the radar receiver experiences a field strength of a third of one-third, or one-ninth, of the original signal. The echo field strength has therefore decreased to a ninth for a threefold increase in range, or, in other words, echo field strength is inversely proportional to the square of the distance of the target from the radar.

Remembering that power is proportional to (field strength)^2 we may state the variation of echo strength in terms of power as:

Echo power is inversely proportional to (target distance)^4.

This is one of the fundamental rules governing the performance which a radar will give for a particular target under free-space conditions. It tells us, for example, that under these conditions every time we double the range of a target we reduce the echo power by 2^4 or 16 times, every time we treble the range we reduce the echo power by 3^4 or 81 times. If we consider a practical target being observed as it moves away from say 0 mile to 10 miles, the power of its echo will fall during this period by 40^4 or 2,560,000 times; this gives some idea of the enormous range of echo powers with which the radar has to deal.

We have now dealt with the variations of echo strength for free-space conditions and have already said that the situation is negligibly changed by the atmosphere; it will be shown later (page 68), however, that the situation is changed considerably when the system is operating near the surface of the Earth. In this case, which is the important one for marine radar, it will be seen later (page 81) that as a large target moves away, the echo power at first decreases with the fourth power of the distance and that, later, near the extreme detection range, it decreases much more rapidly; in this region it decreases approximately as the eighth power of the distance. Here doubling the range would decrease the echo power by 256 times; in fact, however, the echoes on a marine radar from most targets when they enter this region, are only about 10 times the power of the weakest detectable echo, and hence are completely lost in the receiver noise long before the distance has been doubled. This region of very rapid decrease (often referred to as the *far zone*) usually occupies only about the last 25 per cent. of the total range over which the target can be detected; it is, however, the important range because the performance in this region determines the maximum range of detection for surface targets. We can see, for example, that if we attempted to double the detection range of a target by increasing the transmitter power we should have to increase it 256 times.

General principles of radiation and coverage diagrams

As has already been stated, the purpose of the aerial is to focus the radiated energy, in much the same way as a searchlight mirror focuses the light, into a narrow beam.

The focusing action of a searchlight is seldom perfect and if a searchlight is turned on to a distant and sufficiently large object, such as a cliff for example, it will usually be found that the circular area of light, instead of being sharply defined, is brightest in the centre and fades gradually to darkness away from the centre. Also, of course, the brightness gradually decreases with distance as the beam spreads. The same thing applies to the beam radiated from an aerial and we may depict the beam from an aerial or from a searchlight diagrammatically as in Fig. 20.

Since there is no sudden edge or end to the beam there is no direct method of stating what the beam-width is, or for that matter what the 'beam-length' is; the designer of an equipment and the user must, however, have some precise means of specifying the width and the general coverage of the beam. The width will be dealt with first.

Radiation diagrams

What is done in practice is to take a cross-section of the beam and measure how the strength of beam or field strength (which corresponds to brightness in the searchlight) varies across it. This can be done by sending out a continuous steady signal from a stationary aerial, setting up a small measuring receiver at a fixed distance and moving it across the beam as illustrated in Fig. 20. If then, the reading of the output of the receiver in volts is plotted on a vertical scale for each position, a curve of variation of field strength with angle at that distance will be produced. (In practice it is more convenient to leave the receiver steady and to rotate the aerial, but the result is the same.) The resulting diagram is known as the *radiation diagram* of the aerial. Sometimes the same diagram is plotted on 'polar' graph paper; it is then referred to as a polar radiation diagram. An example of a radiation diagram is shown plotted in both forms in Figs. 21 and 22. When studying a radiation diagram it must be borne in mind that it represents the variation in field strength across the beam *at a fixed distance from the aerial.*

The beam-width is defined as the width between points where the field strength has fallen to an agreed percentage of the maximum value. For example, in Figs. 21 and 22 the overall beam-width between points at which the voltage is half its maximum is 20°, occasionally written ±10°. Unfortunately, there are two different conventionally accepted values of field strength commonly used in specifying beam-width, 50 per cent. field strength and 71 per cent. field strength; these are often referred to respectively as the quarter-power and half-power points, or as the 6db and 3db points (see Appendix III). As may be seen from Figs. 21 and 22 the choice of reference value materially affects the width of beam specified and therefore, when it is important to be accurate, a beam-width should not be quoted without stating the convention used.

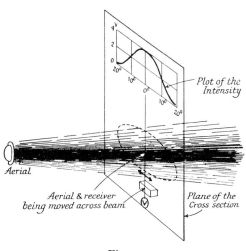

Fig. 20.

In the foregoing, radiation diagrams have been depicted by a curve drawn on a scale of voltage, but it should be remembered that although the *relative* values of voltages are determined solely by the characteristics of the aerial, the *actual* values depend upon the strength of the transmitted signal, the sensitivity of the receiver and the distance at which the test was conducted. There would be no point in publishing a radiation diagram of an aerial in which the values shown depended upon the particular test equipment and distance which someone had happened to use in carrying out a test. This is usually avoided in published radiation diagrams by calling the maximum value of the curve *100 per cent. field strength*, and expressing all the other values of voltage as percentages.

If, in the test described above, the aerial were to be rotated on past the position where the main beam was directed at the receiver it would be found that there were a number of weak subsidiary beams at various angles each side of the main beam as shown in the polar diagram of Fig. 23. These are known as *side-lobes* and are always present in any practical aerial. Since there is a risk that signals given by these may be confused with those given by the main beam, their presence is clearly a nuisance and considerable effort is devoted in aerial development to reducing their strength as

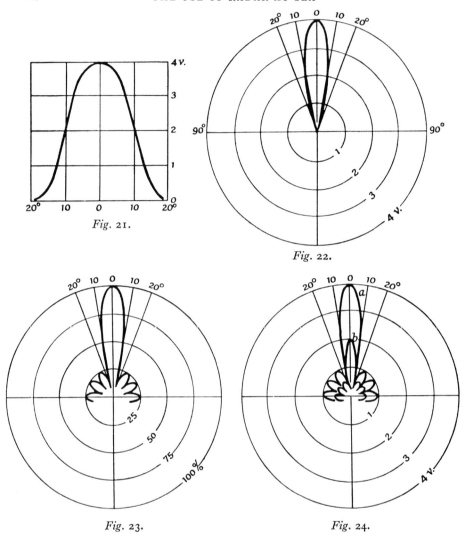

Fig. 21.

Fig. 22.

Fig. 23.

Fig. 24.

Fig. 25.

PROPAGATION OF WAVES AND RESPONSE OF TARGETS   63

much as possible—a figure of the order of 5 per cent. of the field strength of the main beam is usually achieved with present-day aerials.

It should be appreciated that what has been written above about an aerial transmitting energy applies equally well to an aerial receiving energy; the directional properties in the two cases are absolutely identical. In radar applications it is usual to employ a directional aerial for transmitting and also a directional aerial for receiving the echo (very often the same aerial serves both purposes). This results in a considerable improvement in directivity; if the aerial used, as an example, in Fig. 23 is transmitting directly towards an object and is then turned off by 10° it can be seen that the field strength at the object will be halved; if an identical aerial is picking up an echo from the object, the strength of the echo received will be halved again if this aerial is also turned off by 10°, making a total reduction in signal strength of 4 times. Clearly the same argument applies if a common aerial is used for transmission and reception. The same improvement results in the case of side-lobes; the strength of echo received on a side-lobe for an aerial having side-lobes of 5 per cent. would be 5 per cent. of 5 per cent., or 0·25 per cent. of the echo received on the main beam.

Coverage diagrams

In this section we have up till now been considering only a cross-section of the beam (variation of the strength of the beam with angle) at a constant distance. In previous sections we considered how the strength decreased with range. If we combine these two separate ideas we can form a picture of how the signal strength would vary as we change the range *and* the angle; in other words we can then obtain an overall picture of the form of the beam. A change of our mental viewpoint must be made here, and, if confusion is to be avoided, the change must be borne in mind. We have so far been considering the *limits of angle* over which more than a particular strength could be obtained; we are now going to consider the *limits of area* within which more than a particular strength could be obtained—the first quantity has been referred to as the *beam-width* for the particular strength chosen, the second will be referred to as the *coverage* for a particular signal strength.

Let us imagine that we measure cross-sections of the beam at two different distances and plot their radiation diagrams. We will find that both diagrams have the same shape but the field strength for the more distant diagram will be less at each angle than it is for the nearer diagram —if the distances were say 1 mile and 2 miles (i.e. the distance was twice), all the field strengths will be exactly halved. Curve *a* of Fig. 24 represents the polar radiation diagram plotted for the 1-mile position and curve *b* represents the polar radiation diagram for the 2-mile position. It will be observed that whereas a signal strength of 2 volts was obtained at 10° when we were at 1 mile, this signal strength is obtained only at the centre of the beam when we are at 2 miles. This suggests that we could draw an entirely different sort of diagram which would show all the *various ranges and bearings* at which the *signal strength has a constant value*. This is shown in Fig. 25 where the solid line shows all the positions where the field strength is equal to 2 volts.

It will be noticed that this diagram, the *coverage diagram* for 2 volts, has the same shape as the polar radiation diagram but it should be remembered that in fact it represents quite a different idea and is plotted in totally different units. A polar radiation diagram is constructed in terms of voltage and bearing for a constant range; a coverage diagram is constructed in terms of range and bearing for a constant voltage.

In practice, it is not necessary to measure two radiation diagrams at different ranges in order to draw a coverage diagram. It can be shown mathematically that it has the same shape as the polar radiation diagram; hence we could plot the polar radiation diagram at any range, say 2 miles, and draw another diagram of exactly the same shape; then if we observed that, say, the voltage was 2 volts at the centre of the beam, we could call our curve the 2-volt coverage contour, mark the range at which

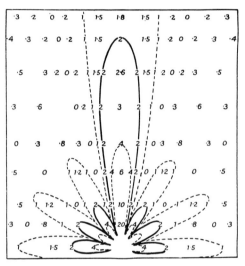

Fig. 26.

it crosses zero angle 2 miles and then divide this 2-mile distance on our diagram up into a suitable range scale. This is what is usually done in practice because it involves the least number of measurements. Another method of obtaining the same diagram, which the reader may find easier to visualize, would be to take readings of field strength over the whole area covered by the beam and to write in the figures in their proper positions on a diagram in the same way that soundings of the sea bed are plotted on a chart, and then to draw a contour line through all the positions of a chosen field strength; such a diagram is shown in Fig. 26 in which the 2-volt contour has been drawn in as a solid line and contours for 1 volt and 4 volts have been dotted.

It should be noted that the contour line of, say, 2 volts cannot be considered as defining the 2-volt coverage *of the aerial*; it is in fact the 2-volt coverage of the particular *transmitter, aerial and receiver* used for the test. It could be expressed in relative units of percentage field strength as was done for the aerial radiation diagrams, but this is usually not done for the coverage diagram because it is, as we shall see later, of such practical value that it is most usefully drawn for a particular equipment with actual values appropriate to the equipment.

PROPAGATION OF WAVES AND RESPONSE OF TARGETS   65

As was said before, the coverage diagram defines the limits of an area within which the signal strength will exceed a chosen value of voltage (2 volts in our example), and outside which the signal strength will be less than this value. This immediately permits a most useful practical interpretation of this diagram. We have shown earlier that the directional properties of the beam from an aerial are the same whether it is a transmitting beam or a receiving beam; consequently if we now imagine Fig. 27 as a receiver coverage diagram, we can see that if we placed a small transmitter 2 miles dead ahead of the directional aerial and if the receiver connected to this aerial is *just* sensitive enough to detect the signal from the transmitter, we could then move the transmitter to any position within the limits of coverage and would still be able to detect it; if we moved it outside the limits of coverage we should fail to detect it; for any position on the curve itself a constant just detectable signal would be received. Similarly, in the two-way radar case, if a target of constant echoing characteristics (i.e. one which always returns a constant proportion of the signal falling upon it) is placed anywhere on this curve it will provide a constant-strength echo if the same aerial (or an identical pair) is used for transmission and reception. This immediately permits a further practical interpretation: if in Fig. 27, for example, curve *a* is taken to be the contour on which echoes from a target of particular echoing strength are just detectable, then it can be seen that the target would be barely detected up to 10° each side at 1 mile and would be detected also on side-lobes at ½ mile. The contour line drawn is for a target having one particular echoing characteristic; the contour for a target of greater echoing strength would, of course, lie outside this contour (e.g. curve *b*) and that for one of smaller echoing strength would lie inside (curve *c*).

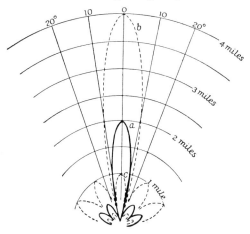

Fig. 27.

It will be clear that an effect similar to that of reducing the echoing strength of the target would be obtained if we reduced the sensitivity of the radar receiver by turning down its gain control. If we do turn down the gain, the size of the coverage area for a given target will also be reduced. If we now imagine that the three curves of Fig. 27 are the contours on which a given target will be just detected at three different gain settings, then the equipment concerned in this figure will first detect this target at 1 mile for a gain setting of *c*, at 2 miles for a gain setting of *a* and at 4 miles for a gain setting of *b*. Similarly, if the target is at a range of 1½ miles it will show a signal over an arc of about 22° with the gain set at b; the arc could be reduced to about 14° if the gain were reduced to a and, of course, if this process is carried too far by reducing the gain to c no signal at all will be observed. Thus, although we cannot alter the actual beam-width of a particular equipment we can have some measure of control of the arc over which a particular strength of echo will paint by altering the receiver gain. In a radar system the arc over which a particular echo will paint will depend upon the strength of the transmitted signals, the position and echoing properties of the target, and the receiver gain; we have no control over the first two of these but a gain control is provided by means of which we can adjust the displayed width of an echo to a reasonable value (and can also reduce the effect of side-lobes).

The fact that a coverage diagram has the same shape as a polar radiation diagram often leads to the two being confused, and it will perhaps be wise at this stage to point out again the difference. A polar radiation diagram is the characteristic of an aerial alone and is plotted in arbitrary units of field strength and bearing; a coverage diagram is a characteristic of a complete system and actual target plotted in real units of range and bearing.

What has so far been written could apply to radiation diagrams and coverage diagrams in both the horizontal and vertical planes; in fact it would apply if the aerial were situated in free space (i.e. well away from the Earth). In practice the aerial is mounted above a curved Earth and this has a profound effect upon the vertical coverage diagram.

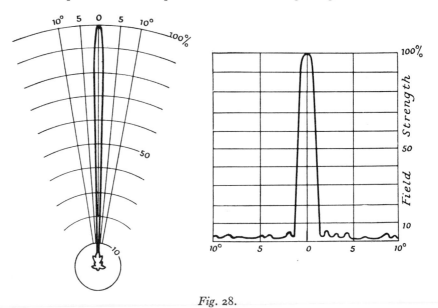

Fig. 28.

Horizontal diagrams

For clarity, the foregoing general diagrams have shown broader beams than would be appropriate to the horizontal case. Horizontal radiation diagrams for an actual marine radar are, however, shown in Fig. 28, drawn in both forms. The horizontal coverage diagram (unlike the vertical coverage diagram dealt with next) is not of any great significance in helping us to understand the performance of the equipment and therefore an actual diagram has not been included; it is generally sufficient to understand how use of the gain control (described above) can be made to sharpen up an excessively wide paint and to reduce side-lobe paints.

The analogy of the searchlight will not help us any further because the beam, being sent out from a circular reflector, is circular in cross-section. The radar beam, on the other hand, has a vertical beam-width of 30–40°, so that the lower part of it strikes the surface of the sea quite close to the aerial. If the aerial were situated in free space, that is, quite clear of the Earth, its radiation and coverage diagrams in the vertical plane would be much the same as those already described generally (Figs. 22 to 27) except that the lobe would be 30–40° wide instead of 20°. However, the position of the aerial in practice is not very high above the Earth, which is a curved conducting surface, and this, as has been said, has a profound effect upon the vertical coverage diagram. The reason for this and its importance will be dealt with in the following sections.

Vertical coverage

The radar horizon. Radio waves have many properties in common with light waves and one of them is that their rays tend to be straight lines. Thus, when operating over a curved Earth, the ultimate limit to the distance at which a target can be detected by radar is set by the distance at which this target falls below the horizon exactly as in the optical case; this would be a straightforward and simple concept if the properties of the atmosphere were constant at all heights.

The navigator will be familiar with the fact that when observing the altitude of a star a correction must be applied due to the fact that the star normally appears to be a little higher than its true altitude and that this correction is greater with low altitudes. In other words the path of the light from the star has been slightly bent by the atmosphere, this bending being caused by the change in the density of the atmosphere with height. The same thing occurs with radio waves but the amount of the bending is rather greater; the path of a radio wave through the atmosphere can therefore be depicted, on a very much exaggerated scale as shown in Fig. 29a. From this figure it may be seen that the horizon for radio waves is a little further away than the horizon for a straight ray.

It is inconvenient to have to think of the paths of radio waves as curved and it would certainly complicate any calculations of distance to the radar horizon. Fortunately this can easily be overcome by the trick of thinking of the Earth as being rather more gently curved than it actually is; we can then draw our diagram as in Fig. 29b. This is equivalent to ascribing a fictitious radius to the Earth, larger than its actual value; experiments have shown that for standard atmospheric conditions, if we adopt a value of

4600 nautical miles, simple calculations based on straight-line paths will give the right answers. Using this value, the familiar expression for horizon distances becomes:

$$\text{Distance to radar horizon} = 1\cdot 22\sqrt{h} \text{ nautical miles}$$

where h is aerial height in feet; this gives ranges 15 per cent. greater than in the geometrical case and 6 per cent. greater than in the normal optical

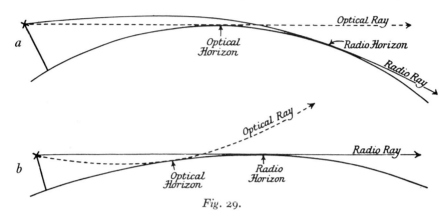

Fig. 29.

case. It must be emphasized that this relation is correct only for *standard atmospheric conditions* (see Chapter 5). For simplicity, the existence of standard atmospheric conditions will be assumed in the following explanations; what happens in other conditions will be dealt with in Chapter 5.

Direct and reflected rays. The vertical beam-width of a radar aerial is considerable and, when the ship is on an even keel, the lower fringe of the beam will strike the surface of the sea all the way from a point quite close to the ship out to the radar horizon. The sea, even when rough, acts as a reflecting sheet which reflects downward-going energy up again, so that a target will receive energy direct from the aerial and also the longer path of reflection from the sea. If it is assumed for the moment that the reflecting sheet is like a flat mirror, the case may be illustrated by Fig. 30. When two

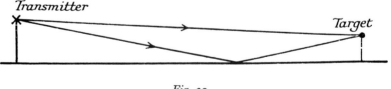

Fig. 30.

parts of the same wave come together after travelling different distances, an effect known as *interference* occurs, which can best be illustrated by an analogy.

Fig. 31 represents waves on the surface of a pond reaching the side in which two channels a and b have been dug. The waves travel with the

PROPAGATION OF WAVES AND RESPONSE OF TARGETS 69

same velocity up each channel but, due to the extra length of channel b, the crest of one wave arrives at the junction at the same time as the trough of the other. The net result is that they cancel and there are no waves in the common channel. In the example chosen, channel b is longer than channel a by half a wavelength, i.e. $0 \cdot 5 \lambda$; the same thing occurs if the difference in path is $1 \cdot 5 \lambda$ or $2 \cdot 5 \lambda$ and so on. If the difference in length is zero, λ, 2λ and so on, the crests and troughs arrive together and the result is a wave in the common channel of twice the amplitude of the wave in either channel.

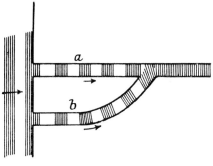

Fig. 31.

The same phenomenon occurs with radio waves and, for the moment, only the outward-going wave will be considered. When the two parts of the wave arrive completely out of step (out of phase) they will cancel one another and when in step (in phase) they will add together. Here there is a departure from the analogy, due to the fact that, on reflection at the surface, that part of the wave is reversed in phase, the equivalent of losing half a wavelength. Thus, the difference in path length which causes the direct and reflected rays to cancel, is an *even* number of half wavelengths instead of an odd number. Precisely the same effect is suffered by the returning echo and subsequent remarks may be taken to refer to the complete radar case.

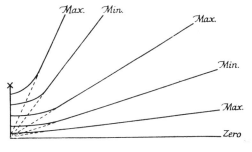

Fig. 32.

Lines of maxima and minima. If the distance between aerial and target (Fig. 30) is increased the difference in path length between the two rays will change. If the target is at a point where the rays cancel and then moves away continuously, the rays will gradually come into phase and out of phase alternately and the resulting signal will pass from minimum to maximum periodically. The same thing will happen if the target height is increased or decreased. This being so, it will be appreciated that if the target is at a point of maximum signal and is moved away, the signal could be kept at a maximum by raising the target at the same time. In fact, as can be shown mathematically, if lines were drawn through all points at which maxima were experienced and at which minima were experienced, a series of curves (hyperbolæ) would be produced as shown in Fig. 32.

The first line of minima lies along the reflecting surface and represents all the points where the path difference is zero, the second minimum is a line along which all the points lie where the path difference is one wavelength, and the third minimum is the line of path difference of two wavelengths. The maxima in turn correspond to path difference 0·5 λ, 1·5 λ, 2·5 λ and so forth. For practical distances away from the transmitter these lines approximate very closely to straight lines originating from a point on the reflecting surface directly below the transmitter.

The effect of this is that the large single lobe which would be the shape of the vertical coverage diagram if the radar were in free space is broken up into a number of lobes by the presence of the reflecting sheet. This is shown in Fig. 33. If the reader finds this difficult to visualize, he should place his hand on edge on the table with the little finger and wrist touching it. If the fingers are parted slightly they will give a very good idea of the lobes of the vertical coverage diagram. It will be remembered that the lines which define the lobes represent the points at which the power has fallen to a particular value.

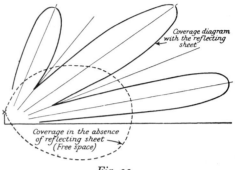

Fig. 33.

Effect of aerial height and wavelength. If the aerial height is increased or the wavelength decreased the angles between the lobes will decrease with a corresponding increase in their number. In other words their number depends on the ratio of aerial height to wavelength. The diagrams show a case where the height of aerial is only a few times the wavelength. With 3-cm. radar and an aerial height of 50 ft. the aerial is approximately 500 wavelengths above the sea. For the flat reflecting surface imagined here the first line of maxima would be at an angle of about 0°·03 and there would be some 500 lobes within the vertical beam-width of an average radar set. Fine as this structure may seem, a simple and very approximate calculation will show that at 10 miles range the lines of maxima nearer the sea surface might be some 150 ft. apart. In reality, therefore, the structure is quite coarse in relation to the heights of practical targets.

Long-range considerations. The assumption of the flat reflecting surface extending from a point beneath the aerial may be used as a fair approximation in practice for very close targets. When a target at longer range is considered, however, it must be imagined that the flat surface is moved away to the point where the reflected ray strikes the sea, the flat plane being at a tangent to the Earth's surface at that point (Fig. 34). This plane will pass beneath the aerial at a distance which may be called the new 'effective height' of the aerial, and the lines of maxima and minima will now originate from the point in that plane nearest to the aerial. The angles between these lines and the flat plane can be calculated from a knowledge of the effective height of the aerial and this would show that they have increased. However, the tilting of the flat plane as it moves away from the aerial will cause the lines of maxima and minima still to curve smoothly downwards in the direction of the Earth's surface. A word of warning is sounded here: in some works on marine radar a simple formula is given relating height above the Earth of points on the first maximum with distance from the transmitter; this formula is valid only at close ranges where the drop due to Earth's curvature is negligible in comparison with aerial height; at longer ranges it gives a height figure which is too small. It is beyond the scope of this book to give the correct formulæ because the computation involved in using them is very large.

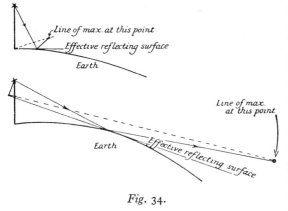

Fig. 34.

A coverage diagram for the curved Earth case may then be expected to take the form shown in Fig. 35 (drawn for the increased radius Earth as in Fig. 29b). The full importance of a coverage diagram of this shape will be dealt with later in this chapter (and a fully calculated typical coverage diagram will also be given) but two points may be noted in passing.

Range and height of first detection. The coverage diagram is regarded as the contour of a just detectable signal from a small target of given echoing strength and if this target is placed in turn at positions *a*, *b* and *c*, it will be strongly detected at *a*, just detected at *b* and not detected at all at *c* although it is still well above the radar horizon. For a target of greater echoing strength we should, of course, draw a larger coverage lobe and it could then perhaps be detected at *c* but there would still be a height at which it would not be detected which would still be above the radar horizon. It can therefore be seen that for a target to be detected it must be appreciably above the radar horizon, the amount above depending upon the characteristics of the equipment and the echoing strength of the target. In the visual case a target may be seen when its top appears above the optical horizon; the radar target, however, has to rise some distance above the

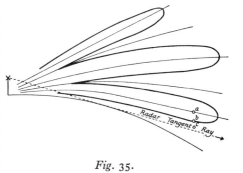

Fig. 35.

radar horizon before it will reach the lowest lobe of the coverage diagram. The sum of the radar horizon distances of the aerial and the target will, therefore, not give the first detection range of any practical target, but only that of an *infinitely strongly* echoing target, a figure which may be substantially greater.

It has been pointed out above that increasing the aerial height or decreasing the wavelength lowers the elevation of the lobes; it can therefore been seen that a radar of short wavelength is at a great advantage over one of a longer wavelength when it comes to detecting low-lying targets. This (as well as the advantage of being able to achieve sharper beam-widths for the same size aerials) was one of the primary reasons for the development of marine radar shifting from 10 cm. to 3 cm.

II—TARGET RESPONSE AND ECHOING CHARACTERISTICS

THE RESPONSE OF SIMPLE TARGETS

When a target is within the radar beam, some fraction of the radio energy directed at it is returned along the original paths. We have previously referred rather loosely to this fraction as the echoing strength or response of the target. The question now arises, on what factors does this fraction depend? As we might expect it depends upon the target size and shape and a little upon the material of which the target is composed.

Material

The echoing properties of targets at the wavelength used for marine radar are not greatly dependent upon the material; the effect of surface roughness is dealt with below under Target Shape. The behaviour is again similar to the optical case: when we see an object it is because of the light reflected or scattered from it back to our eyes; if it is to be impossible to see it, all the light must be absorbed by it (i.e. it is a really dull black) or pass through (i.e. it is completely transparent) or be reflected away. Materials which fulfil any of these conditions completely are no less rare for the radar case than for the optical; certainly the chance of meeting them in the normal course of events is small.

The radar-reflecting properties of metal or water are in fact better than those of wood, stone, sand or earth, but in general the difference is not significant. Unquestionably the effect is small compared with that of the other characteristics of a target such as its size and shape, and for most purposes need not be taken very seriously into consideration.

Target size

It might be expected that the echoing strength of a target would increase with increase of size or, more precisely, with the projected area (i.e. area as seen from the direction of the radar). This is true with certain limitations.

PROPAGATION OF WAVES AND RESPONSE OF TARGETS   73

If a small target is lying in the centre of the beam and is imagined to be gradually widened horizontally, its edges will extend into an ever weakening region of the beam; after a while further increase in size will result in a negligible increase in echo. The target width which can contribute to the strength of the echo is therefore always limited, roughly, to the beamwidth. For example, if the radar were directed straight at a uniform line of cliffs half a mile long, these would fill the major part of the 2° beam of a radar set at 10 miles; if the length of this line of cliffs were suddenly extended to several miles we should not obtain any appreciable increase in echo strength.

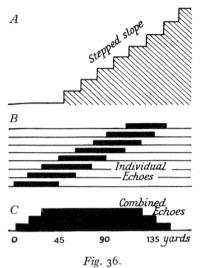

Fig. 36.

The same argument applies to the vertical size of a target, but as almost all targets are small compared with the total vertical beam-width of a marine radar, this limit is not likely to be reached in practice. A limit to the effective vertical extent may very well be set, however, by quite a different factor.

Let us consider a hill face which slopes upwards away from the radar and let us imagine that the face of this hill is composed of a number of steps as shown at A in Fig. 36. Let us further imagine that the horizontal depth of the steps is 15 yards and that the radar is emitting pulses of length equivalent to a distance of 45 yards on the PPI. Now if we were able to see the echoes reflected from each step separately on the PPI they would appear as in B of Fig. 36; in practice, of course, we cannot see them separately; we can only see the combined effect* of the individual strengths of such echoes as are present at any instant as in C of Fig. 36.

It will be noticed that not more than three echoes exist at any one instant (i.e. range), and thus the strength of echo reaches a maximum value after only three steps; in other words, only those steps lying within a pulse-length ($3 \times 15 = 45$ yards) contribute to the strength of echo; further steps lengthen the echo but do not increase the strength. It is emphasized that Fig. 36 at C illustrates the strength and not the shape of the echo.

Exactly the same rule applies to a smoothly sloping surface: only that part of it lying within the pulse-length will contribute to the strength of the echo. If a hill slopes upwards at say 45° (and, with the exception of cliffs and mountain bluffs, few natural slopes exceed this steepness) for

* The radio engineer will realize that the addition will depend upon the phase relations of the individual components; the case is more correctly stated by assuming the faces to be randomly distributed and to arrive at the average value of the signal by statistical evaluation.

hundreds of feet, the strength of the echo will be the same as if the vertical extent were only about 150 ft., assuming a pulse-length equivalent to 50 yards.

To summarize, the echoing strength of a small target varies with its projected area; for a large target a limit is set in azimuth by the beam-width and in the vertical direction by the height interval which corresponds to pulse-length, or by the vertical beam-width if this is less than the height interval (a most unlikely eventuality).

It can now be appreciated that as far as echo strength is concerned, however large the target, it appears to the radar aerial as a number of targets each limited in size to about 2° in azimuth by a few hundred feet or less in vertical extent, the echo strength being equal only to that of the strongest *individual* echo. An optical analogy may help to make this clear. If an observer is standing some distance away from an electric sign of the type that is made up of a number of electric lamps, he might expect that the larger the sign the brighter it would appear to him. However, if he were to view the sign through a telescope so that he saw only one or two lamps at a time it will be clear that increasing the size of the sign would not now result in any increase of brightness; if he swings his telescope he can tell that the sign is larger but he will not say that it is any brighter. In the radar case, increasing the size of a large target will give more extensive but not stronger echoes. The largest mountain may not, therefore, give very strong echoes on a radar which views it piecemeal.

Target shape

Up till now targets have been considered solely in relation to the amount of energy they return *to the radar aerial* as an echo. When a radar beam strikes a target, the energy in that part of the beam which falls on it must be either absorbed by it or re-radiated, the amount of re-radiated energy being determined by target material and size as described above.

It will be realized, however, that not all of the re-radiated energy will reach the radar aerial. If this energy is re-radiated in a diffuse manner over a large angle it will be clear that only a small part of the total energy will reach the aerial; if, however, the energy is re-radiated as a concentrated beam a large part of it may reach the aerial if the beam happens to be aimed back at it. We can represent the width of the angle over which this energy is re-radiated by a target in the form of a polar re-radiation diagram as we did for the initial radiation from the aerial—this will be referred to as the *target polar diagram*; its form is dependent on the target shape.

The sphere

When a radar beam falls on a *small* sphere, the radio energy re-radiated by it is 'scattered' uniformly in all directions, as depicted approximately in Fig. 37. There is no simple explanation of this and the reader is asked to accept it.

PROPAGATION OF WAVES AND RESPONSE OF TARGETS   75

In this case we would expect that only a very small part of the total re-radiated energy would be returned in the direction of the radar receiver, and this is in fact true. The target polar diagram for a sphere would therefore take the simple form of a circle—that is to say that if we measured the field re-radiated from a small sphere it would be constant throughout the whole 360°.

Fig. 37.

From the point of view of echoing strength the sphere is a poor target. However, precisely because of the property that it scatters all the energy uniformly, it is a comparatively easy matter to calculate the strength of signal which a sphere will send back as an echo; for this reason it is an extremely useful standard of reference.

The plane

In considering the echoing strength of a flat sheet or plane we must consider two cases: first of all when the plane is rough and secondly when it is smooth. We can illustrate both cases by an analogy.

Imagine that the white cement wall of a house on the shore is being illuminated by the searchlight of a ship steaming past the house. We will notice first of all that the wall shines back at us quite brightly and that the glass windows appear dark against the wall. As we become more and more at right-angles to the building the brightness of the wall increases very slowly; quite suddenly, however, we get a brilliant flash of light reflected from the windows followed immediately by darkness from them as we pass out of line. The difference in behaviour is due to the fact that the rough walls scatter the light in many directions whereas the polished glass reflects it in a definite direction—this latter effect is referred to as *specular reflection*. Similar effects are found in the radar case, but where we were previously talking of 'rough' and 'smooth' in relation to the minute wavelength of light we must now think of the same terms in relation to the much coarser wavelength of radar. A roughness of $\frac{1}{4}$ inch would be moderately smooth to 3-cm. radar and the concrete wall of our house would certainly be adequately smooth to give specular reflection. Rocks on a seashore, the broken face of a cliff, scrub or trees on a mountain face are what we mean by rough to 3-cm. radar.

The characteristics of rough and smooth planes can be depicted by target polar diagrams as shown in Fig. 38. In these the beam is regarded as horizontal and the plane as vertical.

In A of Fig. 38 the smooth plane has a sharp target polar diagram and when the beam is not at right angles to the plane the re-radiated energy is directed away from the radar receiver. Some of the energy is usually re-radiated as weak side-lobes, however, so that it may be possible to obtain a weak echo even when not at right angles to the plane. When at right angles, of course, almost all the energy falling on the plane is beamed straight back at the receiver and a very strong echo is produced. Within the limits mentioned earlier the larger the plane, the stronger is the echo; however, the larger plane sends back a sharper beam so that to reap the advantage of the stronger echo we must align ourselves more accurately with the perpendicular. Once we are off the line the strength of echo for a given inclination can be shown mathematically to be almost independent of the size of the plane.

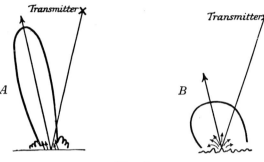

Fig. 38.

Fig. 38 at B shows the target polar diagram of a rough plane. It is a broad beam indicating that the re-radiated energy is being scattered in many directions; it has some directivity indicating that there will be a slight change of echo strength with angle. As we would expect from the scattering, the maximum echo strength will be far less than that from a smooth plane of the same area at its maximum but will be more than that from a sphere of the same projected area. The strength of echo and the directivity will both be dependent on the size of the target and the degree of roughness. This applies whether the plane is inclined in the horizontal or in the vertical direction.

Perpendicular planes

If two smooth vertical planes are arranged at right angles it is possible to achieve the extremely strong reflecting properties of a single plane without the critical dependence on inclination. This is illustrated in Fig. 39. Simple geometry will show that, if the beam is horizontal, for any angle of incidence the angle between the incident and reflected beams ($\alpha+\beta$) will always be 180°, in other words the beam is reflected back parallel to its original course.

An extension of this is to have three planes mutually at right angles; reflection back in the desired direction is then achieved within wide limits of the inclination of the beam in both the azimuth and elevation planes. A specially constructed form of such an arrangement is known as a corner reflector and its use and behaviour are dealt with fully in Chapter 11.

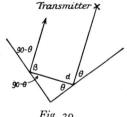

Fig. 39.

The cylinder

The cylinder may be regarded as a combination of the sphere and the plane. If the cylinder is considered to be standing vertically on its end, then its effects in azimuth are the same as those of the sphere; that is to say, it will scatter uniformly in all azimuth directions and consequently tend to provide a poor target. Its effect on a beam which is varied in elevation will be the same as for the plane, that is it will be sharply directional in elevation if smooth and slightly directional if rough; it will thus be a moderately good overall target when viewed perpendicularly to its axis.

The cone

For practical purposes, the cone may be regarded as a tapered cylinder having the same effects in azimuth and elevation as the cylinder except that, when the base of the cone is in the horizontal plane, the reflected beam will tend to be sent upwards, away from the receiver. A smooth cone so placed is unlikely to return any echo to a receiver in a ship. A rough cone will return some, but it will be a poor target.

Target equivalent echoing area

It is instructive to consider the relative echoing strength for the different types of target we have been considering above. Let us take four targets all having the same projected area of 1 sq. ft. The relative echoing properties will then be as in the table below calculated for 3 cm. wavelength.

TABLE III

Target shape	Aspect	Projected area (sq. ft.)	Relative echo power
Sphere	—	1	1
Plane	Perpendicular in both directions	1	1200
Cylinder	Perpendicular	1	300
Corner reflector	—	1	1200

This table emphasizes the way in which the echo strength is extremely dependent upon the shape of the target by showing that, for example, we should need a sphere having a projected area of 1200 sq. ft. (40-ft. diameter) to provide the same strength echo under the same conditions as that given by a 1-sq. ft. plane correctly orientated. This in fact provides a reference against which the *relative* echoing strength of any target may be defined; we say that our sharply reflecting 1-sq. ft. plane gives the same echo as a uniformly scattering sphere of 1200-sq. ft. projected area.

Another way of describing this is to say that the echo strength of the concentrated beam from the small plane is the same as in that part of the total scattered energy from the large sphere which goes in the direction of the radar aerial. We can therefore define the echoing strength of a target in terms of the area of an equivalent sphere and can use the term *equivalent echoing area*. (The expressions, *effective echoing area* and *radar cross-section* are sometimes used for the same term.) This definition is useful because it means that one set of curves specifying the performance of an equipment against a whole variety of different targets can be drawn in terms solely of target equivalent echoing area and these can be used for targets of any shape so long as their equivalent echoing area is known.

The equivalent echoing area of a target is a concept which must be used with care because, for a particular target, its value usually changes with inclination and its value must be specified for the inclination required. For example, if our 1-sq. ft. plane consists of a very thin sheet, its equivalent echoing area is 1200 sq. ft. when viewed perpendicularly and becomes zero if the sheet is viewed end on.

One other word of warning is that the equivalent echoing area of a given shaped target may not vary in direct proportion with its projected area. In the case of the sphere it does; in the case of the plane and the corner reflector it varies as the square of the area; in the case of the cylinder it varies directly as the diameter and as the square of the length. Thus for larger targets, the *ratios* given in the table may be very much increased.

THE BEHAVIOUR OF SIMPLE TARGETS

A typical vertical coverage diagram

Now that a means of defining target echoing characteristics has been established, an actual calculated coverage diagram for a typical marine radar can be given.

Fig. 40 shows the vertical coverage diagram under standard atmospheric conditions out to a range of 30 nautical miles for targets having equivalent echoing areas between 80 and 320,000 sq. ft. Fig. 41 shows an enlarged section of this diagram for the region within radar horizon distance and covers targets of equivalent echoing areas down to 0·3 sq. ft.

The diagram is drawn for calm sea conditions. In a rough sea the minima at close range will tend to be filled in, but the bottom lobe will not be much affected. At radar-horizon range and beyond, the diagram will be substantially unchanged even by very rough seas. It should perhaps be pointed out that, apart from the effect of slight changes in aerial height, the *positions* of the lobes do not change when the ship is rolling, although their size may. It should also be realized that the strength of echoes from practical targets usually fluctuates rapidly to some extent; in practice, a target may therefore be detected intermittently just outside the contour corresponding to its normal size and may also disappear intermittently *just* inside that contour.

PROPAGATION OF WAVES AND RESPONSE OF TARGETS 79

The diagrams are calculated for an aerial height of 50 ft. above sea level and for a radar having the following characteristics.

Wavelength	3 centimetres
Peak power	30 kilowatts
Pulse-length	0·25 microseconds
Horizontal beam-width	2° to half voltage
Vertical beam-width	40° to half voltage
Overall sensitivity	2·75 × 10⁻¹¹ watt

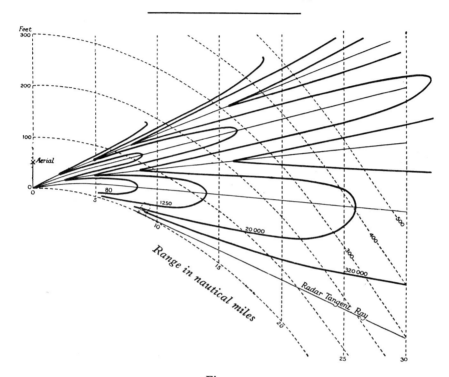

Fig. 40.

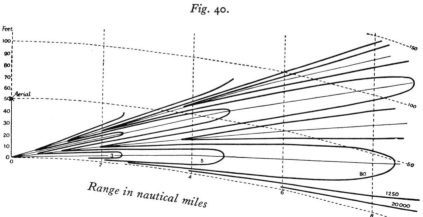

Fig. 41.

The effect of target position

Let us consider a small target of large equivalent echoing area, such as a corner reflector (see Chapter 11), moving in towards a radar at a constant height above sea level. Let us imagine it to have an equivalent echoing area of 20,000 sq. ft. and a height of 50 ft. and follow it on the coverage diagram of Figs. 40 & 41.

At a range of 15 miles it is below the lowest lobe for a 20,000-sq. ft. target and consequently will not be detected. At 14 miles it first touches the 20,000-sq. ft. lobe and hence is first just detected at this range. At a range of about $8\frac{1}{2}$ miles it crosses the first maximum: it will be noticed

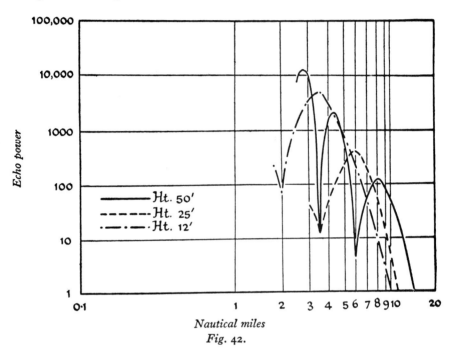

Fig. 42.

that a target having an equivalent echoing area of only 80 sq. ft. could be just detected here so that our corner reflector is now giving an echo power 20,000/80 or 250 times that necessary for detection. With further decreases in range the echo goes through the first minimum at 6 miles, through the second maximum at about $4\frac{1}{4}$ miles where its strength is rather more than 5000 times that required for detection, through a further minimum at about $3\frac{1}{2}$ miles and so forth.

The usual way of displaying the information for this target would be to plot it as a curve of echo strength against range. It is usual to do this on paper having a logarithmic scale; such a diagram for the target we have been considering is shown in Fig. 42. Also shown dotted in this figure are similar curves for an identical target at heights of 25 and 12 ft. These curves do in fact represent precisely the variations of echo strength which

PROPAGATION OF WAVES AND RESPONSE OF TARGETS 81

a radar experiences when approaching or receding from a corner reflector; it must be remembered, of course, that this will not be seen as a change of *brightness* owing to the action of the display in limiting the brightness of echoes at a value only two or three times that of the minimum detectable echo (see Chapter 2).

It will be noticed that the maxima and minima for the three targets of different heights occur at different ranges and that if they approached together towards the radar there would be a tendency to fill in the minima of the combined echo. If we now think of a very large number of small reflectors, or of one continuous plane extending upwards for 50 ft., it is not

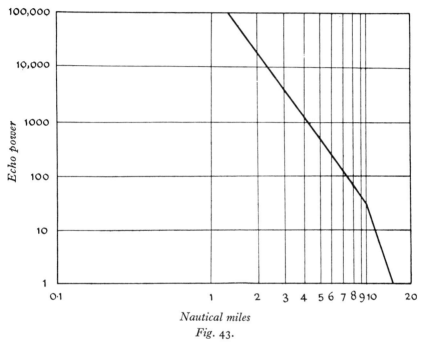

Fig. 43.

difficult to suppose that we should get an echo strength curve of the form shown in Fig. 43. This happens in practice with any target which has a uniform vertical extent comparable with the vertical thickness of a lobe; this is why we do not experience maxima and minima when observing ship targets. When such an extended target is at close range it is illuminated by a number of lobes and as the range is increased the echo decreases at a steady rate until the target has about the same height as the first maximum, when the curve bends over and the echo decreases much more rapidly as the target begins to fall below the lobe. In the first region the power of the echo decreases approximately 16 times each time the range is doubled and in the last region it will decrease at the rate of 256 times each time the range is doubled; these are the regions mentioned on page 51 where the echo strength is inversely proportional to (distance)^4 and (distance)^8.

It is instructive to examine what happens to the echo-strength curves of extended targets when the target equivalent echoing area and the target height are changed.

Fig. 44 shows at *a* the curve of a target of small equivalent echoing area and at *b* the curve of a target of considerably larger equivalent echoing area at the same height. At any range the latter target must give a stronger echo (within the limits of beamwidth) so its curve must be drawn above that of the smaller target. Since both targets are at the same height they pass through the first maximum at the same range and hence the 'knee' of the curve will occur at the same range.

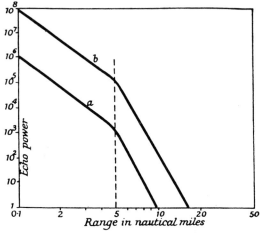

Fig. 44.

Fig. 45 shows the curves of two targets of the same equivalent echoing area at different heights. At close range increasing the height of the target makes practically no difference to the echo strength and hence the curves of the two targets are substantially coincident. The higher target *d*, however, will pass through the first maximum at considerably greater range than the lower target *c* and hence the 'knee' of the curve occurs at longer ranges than that for target *c*.

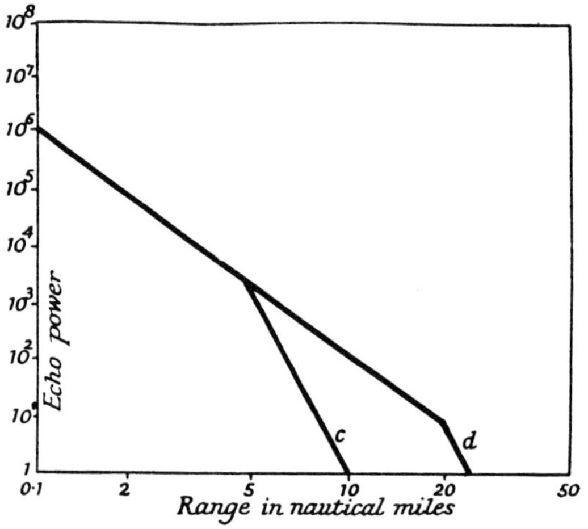

Fig. 46 shows target *b* of the first example (the low target of large equivalent echoing area) and target *d* of the second example (a high target of small equivalent echoing area). It is immediately apparent that the target which was giving the weaker echo at close range continues to give an echo at long range *after the stronger echo has completely disappeared*. An optical analogy may help in visualizing this. When close to a shore, a car headlamp on the foreshore may be much brighter than the lighted window of a house on a cliff top, but at greater range the lighted window will still be seen when the car's headlamp is below the horizon. This effect is almost

Fig. 45.

PROPAGATION OF WAVES AND RESPONSE OF TARGETS 83

certainly never experienced with ship targets because as a ship's structure rises from sea level, increase of height must also result in increase of equivalent echoing area; a weak echo from a considerable height cannot therefore be due to a ship target. This can, and does, occur with land targets, however, due to the radar's property of looking separately at horizontal slices of a hill. Individual structures on a hillside may also provide strong echoes.

RESPONSE AND BEHAVIOUR OF PRACTICAL TARGETS*

Buoys

Navigational buoys usually provide poor radar targets because of their low height, small size and unfavourable shape. The pillar buoy is at some advantage over most other types of buoy because of the height of its central structure. The conical buoy is a particularly poor target because its inclined face causes specular reflection in an unfavourable direction.

Ships

The target polar diagram of a ship is, as might be expected from the complex structure, a very involved one consisting of a large number of lobes. It is thus to be expected that the strength of the echo from a ship will fluctuate considerably with changing aspect; in fact it does fluctuate with hardly appreciable differences of inclination. The shape of the polar diagram will vary from ship to ship but it can generally be assumed that a ship will provide a much stronger echo when beam-on than when end-on. The equivalent echoing area bears a fairly good relation to tonnage and therefore detection range can be used as a fair indication of the size of the vessel (exceptions must, however, be expected in some cases). In the case of freighters and tankers, however, the extent to which they are laden has an appreciable effect. Most of the echoing area is accounted for by the hull and superstructure, the masts and rigging adding very little to it.

Open fishing vessels, yachts and heavily laden barges are usually extremely poor targets.

Icebergs

Large icebergs with nearly vertical faces will usually have a large equivalent echoing area and are likely to be good targets.

Icebergs with sloping faces may give specular reflection (see page 75) in an unfavourable direction and thus have a very small equivalent echoing area even though their actual size is large. Such icebergs will be very poor targets. Icebergs which have some faces nearly vertical and other faces sloping will change from good targets to very poor ones as the aspect changes. Growlers will be very poor targets because of their small size and their shape.

* Additional information will be found in Chapter 6.

Sea clutter

In a calm sea very little sea echo is received by the radar, as almost complete specular reflection of the beam away from the aerial occurs. When there are waves on the surface, however, while some of the energy is still specularly reflected some of it at close ranges is scattered by them and some of it returns as sea clutter; at longer ranges, where the angle of incidence is smaller, more of the energy is specularly reflected despite the presence of quite large waves.

Because of the fact that the scattering is less at longer ranges, the decrease in the strength of sea clutter with range is more rapid than for other types of target; the decrease

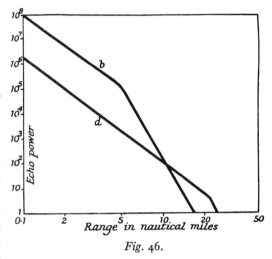

Fig. 46.

follows a known mathematical law, however, and use is made of this fact in the swept gain anti-clutter device dealt with elsewhere.

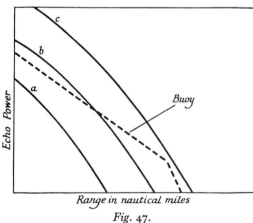

Fig. 47.

Curves *a*, *b* and *c* in Fig. 47 show echo-strength curves of the sea clutter for various states of sea, and the curve of the echo strength of a buoy for comparison. In case *a*, the buoy could be readily detected if just sufficient swept gain were used to remove the sea clutter from the display. In case *b* the buoy echo could be seen at longer ranges but at closer ranges it is buried below the strength of the sea clutter and would also be removed by swept gain. In case *c* it is buried at all ranges.

Chimneys, gasholders and lighthouses

Chimneys and gasholders of *cylindrical* shape usually provide good targets, the gasholders particularly so in view of their large size. Chimneys tapered (or conical) in shape may provide only very poor signals: it must be remembered that the cylinder is a poor target considered in the horizontal plane alone; it becomes a reasonably good overall target only because of its highly directional properties in the vertical plane. If the face is inclined to the vertical by even a very small angle, as it is in the tapered chimney, the directional properties are of no advantage and the chimney becomes a very poor target. Almost all isolated lighthouses are of substantially conical form and therefore provide extremely poor targets.

Buildings

Many buildings, particularly modern factories, behave chiefly as large specular reflectors and can therefore give powerful echoes when perpendicular to the beam. Two buildings at right angles often act with the ground as a huge corner reflector giving an extremely powerful echo over a wide variation of aspect. Built-up areas are the source of the strongest echoes known in the whole of the radar field.

Land targets

Most mudbanks, sandbanks and foreshores are sufficiently smooth to act mainly as specular reflectors and, if their inclination to the horizontal is small, will provide extremely poor targets; this combined with their low height makes the chance of detecting them very slim, except at very close range. Their surfaces are sometimes sufficiently indented or rippled to cause some scattering and hence weak echoes. Vegetation above high-water level or a pronounced back to the beach will provide scattering and hence a better, but still poor, target. A vertical sea wall when viewed at right angles will provide a comparatively good target, but if its height is not very great it will not be detected at very long range.

Cliffs provide very favourable targets. Their faces are usually sufficiently broken to prevent complete specular reflection but not broken enough to provide very wide angle scattering. The result is a horizontal target polar diagram which has a fair degree of directivity but is not excessively sharp.

Cliffs are nearly vertical, so that the radar aerial is very near the maximum of the vertical scattering diagram and, as long as it is not too far from the perpendicular in the azimuth direction, we should expect the cliffs to behave as very good targets. The fact that the face is nearly vertical means also that the whole of its possibly several hundreds of feet of height will lie within a pulse-length and can contribute to the signal strength. The height of cliffs will tend to bring them above the radar horizon of the aerial at great distances, and so everything conduces to the probability of their being detected at long ranges.

Hills and mountains with comparatively gentle slopes (say 45° and less) tend to fare worse than cliffs. If the slope acted as a specular reflector it would return no signal; if it is to be detected it must act as a scattering target (which it is likely to be in fact); its directivity will be low, which implies a poorish target. The property of the radar of viewing a large hill or mountain piecemeal prevents the *size* of a hill being a factor of importance. The strongest sources of echoes may very well be provided by comparatively small man-made structures on the hillside or by a flat rock face momentarily having a favourable aspect.

One other factor in the behaviour of a high hill or mountain is of considerable importance. For the reasons given above, a hill must be considered not as a continuous target but as a large group of independent targets. If the hill or mountain is at long range, the vertical extent of any one of these targets will be small compared with the vertical thickness of the lobes of the coverage diagram. Therefore a favourable target which is in the maximum of a lobe at one range may be in a minimum at a closer range. This means that as we change the range, the parts of the hill giving the strongest echoes will change very markedly, rather as the reflection of the sun from a number of tin cans on a scrap heap dances about in a random manner as we change our position. Combine this with the fact that the echoes will change with aspect (i.e. in the horizontal plane) and it will be seen that the curve of variation of maximum echo strength with range will be an exceedingly complex one, and there is no reason why the curve for one mountain should resemble that for another.

It can be seen from the foregoing that if a mass of echoes is being received from a range of hills, the identification of a feature expected to give the strongest echo is fraught with great difficulty. It is perhaps not impossibly difficult but it demands considerable experience, observation over a change of range, and a sufficiently open mind to recognize the possibility of being wrong.

What of practical value can be salvaged from this depressing picture of difficulties? We cannot readily predict which hill tops or mountain peaks will be detected at a particular range but we can predict, for standard propagation conditions, which ones *cannot*. A knowledge of the heights of the peaks will show which are likely to be below the radar horizon and therefore will not be detected. This will reduce the number of uncertainties and probably be of considerable assistance.

To summarize the case of mountain echoes we can say that almost the only feature in their favour is their height. Increase of height increases the *possibility* of detection at long range but *not its certainty*. At a given range there is no reason why the higher mountain need give the stronger echo. Where a high cliff is surmounted by a gently rising hill, at a given range the cliff will *probably* give a stronger echo than the hill; when closing from a long range the cliff may *possibly* be detected even before the higher hill.

Some typical performance data

Measurements have been made of the echo-strength curves of various targets as obtained by a typical marine radar of the same characteristics as those specified in connection with Figs. 40 and 41. These are shown in Fig. 48. All the curves for ships are for the end-on aspect.

In view of the complex story outlined above, there would be little point in obtaining echo-strength curves for land targets—indeed the task of measuring them would be almost impossibly difficult. Fig. 49 may be of some value, however. It shows plots of a number of reports of detection ranges achieved on land targets of various heights and a band about the mean of these plots. Also shown on this diagram is a curve showing the

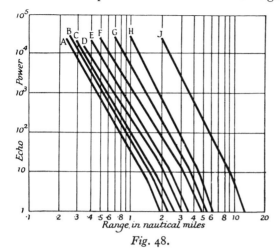

Curve:

A. 30-ft. boat
B. 3rd-class buoy
C. 2nd-class buoy
D. Fishing boat (70 ft.)
E. Pillar buoy
F. 100-ft. boat
G. 200-ton ship
H. 1000-ton ship
J. 10,000-ton ship

Fig. 48.

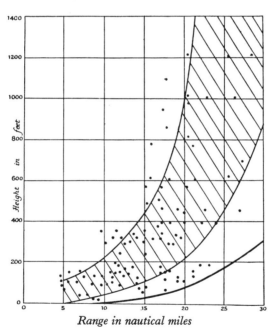

Fig. 49.

ranges at which targets of various heights will begin to appear above the radar horizon for an aerial height of 50 ft. For a land target of given height under standard propagation conditions it can then be said that it cannot be detected for any position to the right of the 50-ft. curve and that it will *probably* first be detected at some position within the shaded band.

5

RADAR METEOROLOGY

THE previous chapter was concerned with the propagation of waves under standard meteorological conditions. Although it is necessary to define such conditions as a basis for calculation and comparison, it should not be assumed that they will be the normal conditions for any particular area. In fact, in many parts of the world conditions will depart from the standard more often than not. The purpose of this chapter is to explain the effects of non-standard meteorological conditions and to deal with the effects of the atmosphere which reduce radar detection ranges and produce 'weather echoes,' i.e. echoes from clouds, rain and other forms of precipitation and from atmospheric discontinuities.

Standard propagation

Any definition of a standard atmosphere is bound to be arbitrary. In this book the definition adopted is as follows:

Barometric pressure at sea level 1013 millibars, decreasing with height at a rate of 36 millibars per 1000 ft.
Temperature at sea level 59° F. (15° C.), decreasing with height at a rate of 3·6° F. (2° C.) per 1000 ft. (305 m.).
Relative humidity 60 per cent. (remaining constant with height).

This definition has been accepted in some quarters, though it is not universally agreed. The values are mean values taken for the whole world and are a useful guide, but it should be again stated that they should not be taken as the normal conditions.

Under such conditions the refractive index of the atmosphere, which is dependent upon all of them, has a value of 1·000325 at sea level and it decreases uniformly with height by 0·000013 per 1000 ft. This is illustrated in Fig. 50.*

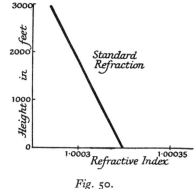

Fig. 50.

In standard atmospheric conditions, because the refractive index decreases uniformly with height, radar rays will be bent into a slight downward curve towards the Earth's surface, with the result that the distance to the radar horizon will slightly

---

* This and the following curves of refractive index must not be confused with the curves of 'Modified Refractive Index' or *M* curves, familiar to the skilled radar meteorologist.

exceed that to the optical horizon, as was explained in the previous chapter (Fig. 29).

Non-standard propagation

Of the factors mentioned above, upon which the rate and direction of the change of refractive index depend, the alteration of relative humidity and temperature with height are the most significant. Propagation conditions may depart from the standard in either direction, causing less or more than the standard rate of decrease of the refractive index with height.

If the refractive index decreases with height at less than the standard rate (i.e. at sub-standard rate) or if it should increase, the rays will be bent less and there will be less tendency for them to follow the curve of the Earth. Hence, the lobes of the vertical coverage diagram will tend to lift further above the surface and the radar horizon will occur at a shorter range. This condition is referred to as *sub-refraction.*

If the refractive index decreases with height at more than the standard rate (i.e. at super-standard rate) the rays will be bent more severely and they will follow the curve of the Earth more closely. In this case the lobes of the vertical coverage diagram will move down closer to the surface and the distance to the radar horizon will increase. This condition is known as *super-refraction.*

In the more detailed explanation which follows it should be assumed that all decreases or increases mentioned are with respect to an increase of height above sea level. For the phenomena to occur as described, it will be necessary for the conditions specified to exist along the whole path between radar and target. If the conditions necessary to produce unusual results are prevalent only in the immediate vicinity of the radar aerial, then it is unlikely that they will cause any appreciable effect upon the propagation of the radar wave.

Sub-refraction

A reduction in the rate of decrease of the refractive index may be caused by the decrease of temperature being more rapid than standard or by the relative humidity increasing. If these departures from standard are sufficient the refractive index will even increase with height.

Fig. 51 shows possible variations of the refractive index under sub-refraction conditions. The solid curve shows the refractive index remaining constant throughout the first 500 ft. above sea level and thereafter decreasing at the standard rate. The dotted curve *a* shows the refractive index falling less rapidly than in the standard condition, while *b* shows the refractive index rising during the first 500 ft. In all three cases radar rays would suffer sub-refraction below that height and in the last case sub-refraction would be severe.

Under these conditions, if the radar rays were depicted passing over an Earth having the increased effective radius which was used in Chapter 4 when considering standard refractive conditions, they would appear to curve upwards. The greater the departure from standard conditions, the greater would be the amount of upward curving. The upward bending of the radar rays under sub-refraction conditions will have the effect of slightly lifting the lobes of the coverage diagram; targets of low height may therefore be detected at somewhat shorter ranges than under standard conditions. There have been instances where the detection ranges of small ships have been decreased by as much as 30–40 per cent.

Super-refraction

Super-refraction conditions exist when the rate of decrease of the refractive index is greater than standard; this may be caused by relative humidity decreasing instead of remaining constant or by the temperature decreasing at a rate less than the standard (under some conditions the temperature may even be found to increase with height). A decreasing relative humidity is the more usual cause of super-refraction.

Fig. 52 illustrates possible variations of the rate of change of the refrac-

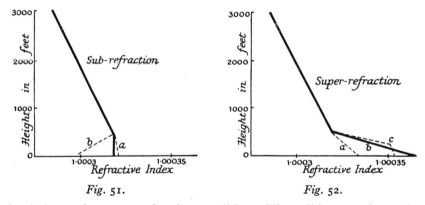

Fig. 51.   Fig. 52.

tive index under super-refraction conditions. The solid curve shows the refractive index decreasing at a much greater rate than standard up to 500 ft. and then taking up the standard rate of decrease. Under this condition, severe super-refraction would occur and radar rays within this part of the atmosphere would be bent down more than under standard conditions and would tend to follow the curvature of the Earth very closely. Curve *a* illustrates a condition in which there would be slight super-refraction and so only a slight increase in the amount of downward bending.

Ducts. If the rays are bent down much more sharply, they may strike the surface of the sea, be reflected upward, again curved down to the sea and so on continuously, thus effectively following the curvature of the Earth and allowing detection at very great ranges. This extreme case of super-refraction is known as *ducting*, the rays being confined within a duct.

Curve *c* of Fig. 52 shows the refractive index decreasing at the standard rate up to 250 ft. then sharply changing its rate so as to decrease much more rapidly up to 500 ft., thereafter resuming the standard rate. In this case, radar rays between sea level and 250 ft. will be curved slightly downward in the normal manner; those which pass above the 250-ft. level will be subject to strong super-refraction and will be bent downwards; they will be reflected upwards again from the discontinuity at 250 ft. (the plane at which the severe change of refractive index occurs), only to be bent downwards once more and so follow a 'hopping' path round the Earth's surface but elevated above it; rays which despite this bending escape into the standard atmosphere above 500 ft. will revert to the slightly curved path which that condition causes.

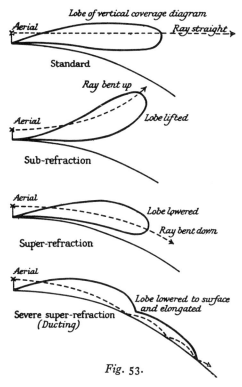

Fig. 53.

A space, such as that in the example, between the 250-ft. and 500-ft. levels, in which the rays may travel for considerable distances following the Earth's curvature is known as an *elevated radar duct*. It is unlikely that elevated ducts will be of much significance to the user of marine radar, since the amount of energy trapped within the duct usually represents only a very small part of the total energy leaving the radar scanner.

Ducting is usually associated with a sharp decrease of moisture content which is often accompanied by a temperature inversion, that is, a condition in which there is a sudden increase of temperature.

Fig. 53 illustrates broadly the effects of standard and non-standard propagation, including ducting (*cf*. Fig. 29b).

Conditions under which non-standard propagation may be expected

As might be expected, the reliable prediction of propagation conditions is an extremely complex matter because of the large number of factors involved and the amount of data necessary. A skilled meteorologist with a wealth of equipment and information at his disposal may make accurate predictions, but the mariner cannot expect to be able to work on this scale himself. However, knowledge of some of the meteorological conditions likely to affect propagation may be of considerable help to the radar user, by putting him on his guard against unusual performance from his equipment.

In the first place non-standard propagation of either kind usually requires relatively calm conditions. If there are very strong winds or other conditions which cause turbulence in the atmosphere, the air at various heights will be sufficiently mixed to prevent any marked irregularities in the rate of change of the refractive index and hence standard propagation may be expected. In anti-cyclonic weather there is usually a subsidence of warm dry air which, settling over the surface of the sea, may produce super-refraction conditions.

Ocean areas. Super-refraction is often experienced in the region of the trade winds. Except in that region and when near large anti-cyclones, non-standard conditions do not often occur in mid-ocean.

Coastal areas. A common cause of non-standard conditions is a considerable difference between the temperature of the wind and that of the sea. This implies the proximity of land masses, and hence that non-standard propagation is more likely to be encountered in coastal waters or restricted sea areas. It is important, therefore, to be on the lookout for such conditions when making a landfall.

Whenever a cool air is blowing over a relatively warm sea there is likely to be a rapid decrease of temperature with height at the lower levels and hence sub-refraction conditions. This is frequently the case in polar regions and in the vicinity of very cold land masses. Where warm dry air is blowing over a relatively cool sea, super-refraction may be expected due to a decrease of relative humidity with a less than standard fall of temperature with height or with a temperature inversion. These conditions may frequently occur in coastal waters in temperate or tropical zones. In the former they are naturally more likely during the summer months. In areas like the Red Sea, super-refraction conditions may be present more often than not. Although a difference of temperature of only 5–10° F. is sufficient to induce the condition, differences of 25–30° F. are by no means uncommon.

ATMOSPHERIC ATTENUATION AND WEATHER ECHOES

In dealing with the propagation of radio waves, Chapter 4 expressly omitted any consideration of attenuation due to meteorological conditions other than to state that range reductions may be expected. It is necessary now to examine how and in what degree this expectation may be fulfilled and what indications may be observed upon the radar display. Attenuation is caused by absorption of energy by the atmosphere and by the absorption and scattering of energy by the various forms of precipitation, rain, &c. These need to be considered separately. The amount of attenuation caused by each of the various factors depends to a substantial degree on the radar wavelength. Consideration here will be confined to the wavelength most commonly used in marine radar, 3·2 cm.

Attenuation by atmospheric gases
Energy is absorbed by the oxygen and the water vapour in the atmosphere. A clear distinction must be made between water vapour and fog. In the

latter, water is held in liquid suspension and not as a gas, and this case is considered separately below. Atmospheric gases cause appreciable attenuation to radio waves but their effect at a wavelength of 3·2 cm. is not great enough to warrant further consideration here.

Attenuation by fog, rain, hail and snow

Generally speaking the amount of attenuation caused by these phenomena is dependent upon the amount of water, liquid or frozen, present in a unit volume of air and upon the temperature. As would be expected, therefore, the effects differ widely, attenuation due to fog being one extreme and that due to tropical rain the other.

It will be appreciated that the further the radar wave and the returning echo have to travel through the absorbing or scattering medium the greater will be the attenuation and the more will the detection range or the echo strength at a given range be reduced. This is the case whether the target is in or beyond the fog or precipitation. When the target is inside an area of precipitation from which echoes are being received, a further reduction of detection range may be expected and this is discussed below under the various kinds of precipitation.

Fog. It can be demonstrated that the attenuation of radio waves in fog may be related to the range of visibility. That is to say, the lower the visibility the greater the water content of the fog and the greater the reduction in detection range. There is some reason to believe that polar fogs (at 32° F.) cause an appreciably greater reduction in detection range than fogs in temperate regions (60° F.), which in turn cause more than those in tropical regions (at 77–86° F.).

In Fig. 54 the reduction of detection range due to attenuation by fog is illustrated for the three regions and for four different types of target. Radar detection range is plotted against visibility. It will be noted that when the visibility is 100 yards or more the detection range of even the best target is only very slightly affected, but that when the visibility is reduced to 50 yards the good target suffers a considerable reduction in detection range while the smaller targets are progressively less affected. The curves are drawn for a radar set with the characteristics given on page 70 and for conditions in which the fog is continuous between the radar and the target.

As will be realized, this data concerns only fogs which are a direct result of the meteorological conditions and takes no account of the presence of matter in the atmosphere other than water. There is very little detailed information available on the effect of the various kinds of smoke &c. associated with fog, or on the effects of sandstorms or sand haze. Numerous reports of sandstorms over the Persian Gulf indicate that their effects are similar to that of fog, although they may be a little more severe for the same visibility. The particles of water contained in fog are so small that the energy scattered back by them in the direction of the radar is too weak to appear as echoes on the PPI. There have been reports of faint indications but this is not the general experience.

Clouds. Owing to the considerable vertical beam-width of marine radar, distant clouds may often be within it. Experience has shown, however, that relatively few clouds will give detectable echoes, unless precipitation from them or within them is occurring. In the latter case no precipitation may be visible outside the cloud.

Rainfall. The attenuation caused by rain is usually sufficient to reduce the detection ranges of targets. It is possible to relate the amount of attenuation to the rate of precipitation. It is dependent also on temperature, but on the wavelength of 3·2 cm., the variation is not large enough to justify considering the three regions separately and the effects illustrated in Fig.

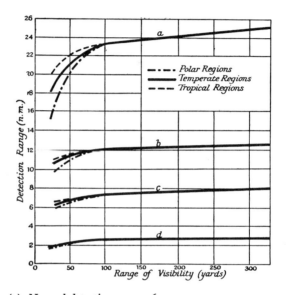

(a) *Normal detection range* 26 *n.m.*
(b) *Normal detection range* 12½ *n.m.* (10,000-*ton ship*)
(c) *Normal detection range* 7½ *n.m.* (1000-*ton ship*)
(d) *Normal detection range* 2½ *n.m.* (*small boat or buoy*)

Fig. 54. *Detection ranges in fog.*

55 may be regarded as applying to all of them. The solid curves in the illustration show the reduction of detection range for various rates of rainfall, *due to attenuation only*, and for three classes of target. They are based on the assumption that the area of rainfall extends uniformly from the radar aerial to *just short* of the target.

If the target is *within* the area of rainfall, any echoes from the raindrops will further decrease its detection range. The drop size of rainfall is such that any but the very lightest rain will scatter sufficient energy back to the radar to produce clutter on the PPI; a limit to the range of detection will be set at the point at which the target echo and the clutter are of equal intensity. The dotted curves in Fig. 55 represent the reduced detection ranges of the three types of target when within the rainfall area.

It will be seen from the curves that the detection range of a 10,000-ton ship, which would normally be detected at 12½ miles, would be reduced by a rainfall of 1 inch per hour to 7 miles if she were just beyond the storm area and to 5¼ miles if she were within it. It will also be seen that the effect on the echoes of the 1000-ton ship and the small boat or buoy when they are within the storm area is very much more pronounced. For ships considerably larger than 10,000 tons the solid curve would be drawn above curve a; the strength of echoes from such ships will be so great that only the heaviest rain will cause clutter strong enough to affect the detection range and no dotted curve would appear.

It must be emphasized that these curves should be used only as a general

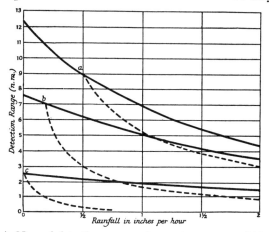

(a) Normal detection range 12½ n.m. (10,000-ton ship)
(b) Normal detection range 7½ n.m. (1000-ton ship)
(c) Normal detection range 2½ n.m. (small boat or buoy)

TARGET JUST BEYOND STORM AREA ―――――
TARGET IN STORM AREA ― ― ― ― ―

Fig. 55. Detection ranges in rain.

guide. They may be regarded as somewhat pessimistic, since it is seldom that the rainfall over a large area will be completely uniform; this is particularly so in the case of very heavy rainfall, in which the storm area is usually relatively small. Further, manipulation of gain, differentiator and circular polarization, if available, may render the solid echo of a target perceptible through the clutter at ranges greater than those indicated.

Fig. 55 may be of more practical interest when it is related to the duration and the frequency of occurrence of rainfall at various rates. Tables IV and V contain statistics for temperate and tropical-equatorial regions.

Hail and snow. The attenuation caused by either hail or snow is very much less than that due to rainfall at an equivalent rate of precipitation (that is, the same quantity of water, when melted, per unit volume). Further, it is extremely uncommon for hail or snow to reach the rates of precipitation frequently experienced in rain. In general, attenuation in hail will be very much less than that in rain at an equivalent rate of precipitation. The reduction in detection range *due to attenuation* in hail and snow is shown in the solid curves of Fig. 56 and a comparison with Fig. 55 will emphasize how small it is. If hailstones are very large (more than $\frac{3}{4}$ inch in diameter) the reduction will be rather greater. As in the case of rainfall, the solid curves represent circumstances in which the precipitation uniformly covers the area around the radar set, the target being just outside this area. On the other hand, *echoes* from hail may be as strong as echoes from rain at the same rate of (water) precipitation. They are only likely to exceed those from rain if the hailstones are more than $\frac{1}{4}$ inch in diameter, which will be unusual.

TABLE IV

Rainfall in Temperate Regions (Temperature 60° F.)

Rate of precipitation (inches per hour)	Duration of rain	Frequency of occurrence of rain (times per year)
$\frac{1}{4}$	1 hr.	At least 12
$\frac{1}{2}$	30 min.	At least 4
$\frac{1}{2}$	15 min.	At least 12
$\frac{3}{4}$	15 min.	At least 4
1	30 min.	About 1
1	10 min.	At least 4
$1\frac{1}{4}$	15 min.	About 1

TABLE V

Rainfall in Tropical-Equatorial Regions (Temperature 75–85° F.)

Rate of precipitation (inches per hour)	Duration and frequency of occurrence of rain
$\frac{1}{4}$	Several hours: very often.
$\frac{1}{2}$	More than an hour: frequently, especially in the rainy season.
2	30 minutes: at least 12 times per year.
$\frac{1}{2}$–2	Rates of rainfall intermediate between $\frac{1}{2}$ inch and 2 inches per hour, and lasting for at least 30 minutes will occur often in the rainy season.
4	Up to 1 or 2 hours: exceptional—perhaps about once per year.

The strength of echoes from snow depends very much on the size of the snowflake as well as on the precipitation rate. In cold climates the snowflake often consists of a single crystal, while in warmer regions it usually comprises a number of crystals, and so is much larger. The echoes obtained from a snowfall of single crystals may be equal to or somewhat stronger than those obtained from rain at the same rate of (water) precipitation. A snowfall of the larger snowflakes may, therefore, produce echoes considerably stronger than rainfall at the same rate. For practical purposes, however, the significant factor is the rate of precipitation, because the water content of the heaviest snowfall will very rarely equal that of even moderate rain. For example, to equal the rate of (water) precipitation of a rainfall of ½ inch per hour, a not unusual phenomenon (Table IV, above), snow would need to increase in thickness on the ground at a rate of 5 inches per hour, which is certainly not common.

The general inference to be drawn is that echoes from snow, particularly in temperate regions, will usually be considerably less troublesome than echoes from rain; in polar regions, although snow frequently falls in the

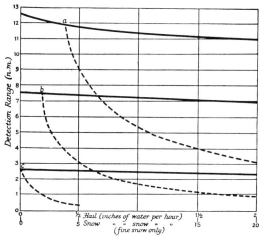

(a) Normal detection range 12½ n.m. (10,000-ton ship)
(b) Normal detection range 7½ n.m. (1000-ton ship)
(c) Normal detection range 2½ n.m. (small boat or buoy)

TARGET BEYOND STORM AREA ————
TARGET IN STORM AREA — — — — —

Fig. 56. *Detection ranges in hail and snow.*

form of single crystals, it may occasionally reach such a high rate of precipitation that its echoes will be more troublesome. The dotted curves of Fig. 56 indicate the reduction in detection range which may be experienced in hail and snow storms of various intensities, when the target is within the precipitation area. In view of the wide variation possible in the size of snowflakes and their differing effects for the same rate of precipitation, it will be realized that for snow conditions the curves should be regarded only as very rough indication. It should also be remembered that the average snowfall in temperate regions will seldom reach the equivalent of ½ inch of rain per hour.

Atmospheric discontinuities, or irregularities in the lapse rate of temperature or humidity in the atmosphere. Echoes caused thereby are mentioned in Chapter 7.

6

INTERPRETATION OF THE DISPLAY

THE picture shown on the PPI of a radar set that has a high degree of resolution and if properly adjusted is a fascinating sight. At a casual glance the detail appears to be amazing and the variety of echoes presented infinite. The close observer, however, will find that, without experience and understanding of the response of targets and the distortions of the presentation, he will not be able to reconcile to his full satisfaction the echoes which he sees with the targets which have caused them.

Echoes of certain coastlines and structures may show striking likenesses to their true forms, but the identification of many familiar objects of navigational interest may prove difficult without experience in observation and skill in operating the controls. The information given in this chapter is intended to assist the observer in identifying echoes with targets, while the following chapter deals with the variety of effects, due to internal and external causes, which tend to increase the difficulties of interpretation.

The PPI picture is a two-dimensional plan in monochrome (usually amber but occasionally green). It is characterized by an extremely limited range of tones. Whereas in a good photograph there is an infinity of gradations between black and white, in the radar display there are only two or three steps between the brightest echo and obscurity. An object which appears substantial to the eye will not necessarily produce a strong echo; as has been mentioned in Chapter 4, the presence or absence of echoes depends on such factors as the height, slope, composition, aspect and distance of the targets.

The size of the PPI picture is a factor which greatly affects the interpretation and the information which can be deduced from it. For best results the diameter should not be less than 12 inches. Many prefer 16 inches, particularly when a reflection plotter is fitted over it.

When seeking identity in outline between the plan view of a target and its echo, the same factors have to be considered in relation to each part of the target which is in the view of the scanner. Further, the distortion of the echo due to the imperfect resolution of the radar set as described in Chapter 1 has an important bearing on echo shape. Detailed consideration of these matters is given in this and subsequent chapters.

Meanwhile it may be useful to describe the appearance of a typical PPI picture. Assuming that the ship is a few miles off shore in a slightly choppy sea the general appearance of the display is likely to be very much as follows. In the immediate vicinity of the small bright spot, which represents the ship herself, will be the indications of sea clutter, which can of course be diminished or eliminated by the use of the sea-clutter control. Beyond a certain range from the ship all these echoes will have faded away,

INTERPRETATION OF THE DISPLAY 99

and the sea surface will be represented by a dark background, against which will stand out the echoes of the coastline and of ships and other objects above water. The coastline may or may not be clearly delineated; beyond it will be seen echoes from inland objects, such as buildings, hills and mountains, while at still greater ranges the screen will again be dark, either because returned echoes from high land will be too weak at this range to paint, or because distant targets are obscured by intervening land. A typical picture is shown in Plate 8.

In Chapter 4 the reader has already been given a full description of the meaning of 'response' and of the factors which affect it. Much practical information has, however, been gathered in the years during which radar has been at sea and the data collected can assist the navigator in making useful forecasts of what he may expect to see on the PPI in particular circumstances and in interpreting the echoes which do appear.

The behaviour of echoes as the ship approaches or passes targets differs very much according to their nature and it may best be considered by examining the different classes of target. For convenience these are classified as Natural Targets, Artificial Targets and Isolated Targets.

NATURAL TARGETS

Echoes from land

Distant land echoes, shortly after their first appearance at their maximum range on the radar screen, can usually be distinguished from ship echoes by their greater size and regularity in painting on consecutive revolutions of the scanner (see Plate 9). Once the average detection range of a particular land target has been determined, this can be a useful help to its recognition in future.

In practice the majority of land echoes, except for outstanding hills and mountains inland, will usually be concentrated within 2 or 3 miles of the coastline. If the coastline is composed of vertical cliffs, these will be found to give a thin but strong line echo detectable at good range (Plate 10); however, if the cliffs are sloping they are likely to give a markedly thick echo which will not be seen at such a long range (Plate 11). Sand or shingle foreshores backed by low-lying land will usually be defined by a thin and smoothly drawn line echo detectable only at a medium range (Plate 12); the evenness of this line will distinguish it from a cliff echo, which is likely to have a ragged appearance.

The echoes from high undulating country will change in shape and disposition as the angle of view changes; and when this causes high land to shield more distant targets, which are capable of returning echoes, a similar change in the appearance of the picture may result (Fig. 57 and Plates 13 and 14). The radar shadow of a mountain is often well defined by contrast with echoes appearing on either side of it.

It will be noted that the range of greatest detection of a hill will not

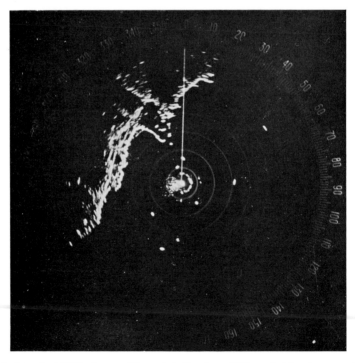

Plate 8. A typical PPI picture. *Normal, indirect, multiple and side echoes are showing.*

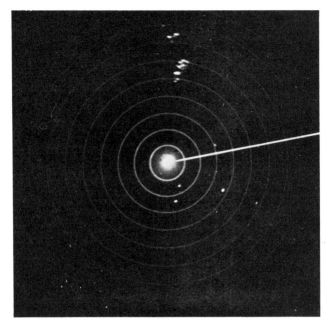

Plate 9. Distant land. *Portland Bill from the southward: range rings 2 miles apart.*

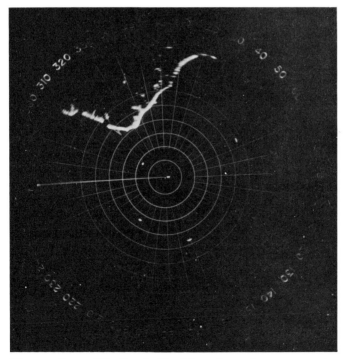

Plate 10. Steep cliffs. *Beachy Head from the south-east: range rings 1 mile apart. (Radial lines here are engraved on the bearing cursor.)*

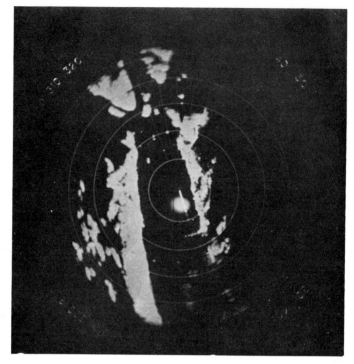

Plate 11. Sloping rock face. *Oslofjord: range rings ½ mile apart.*

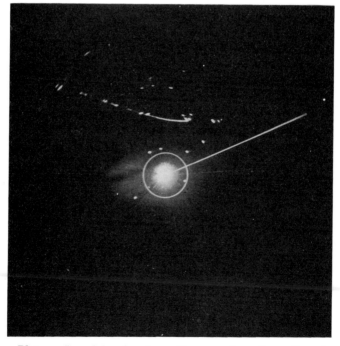

Plate 12. Low-lying foreshore. *Dungeness from the southward.*

INTERPRETATION OF THE DISPLAY 103

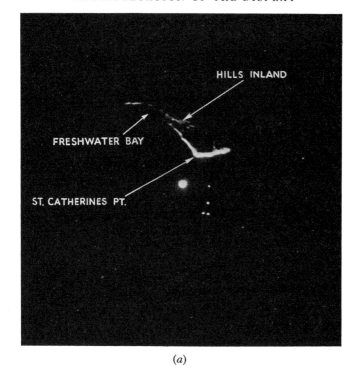

(a)

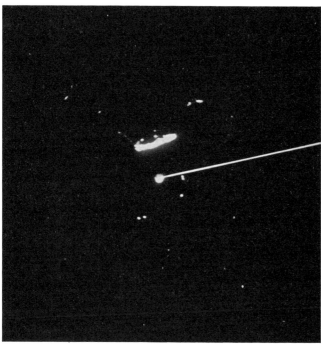

(b)

Plate 13. Change of appearance with angle of view. *In (b) the ship is 1 mile further east and Freshwater Bay is shielded.*

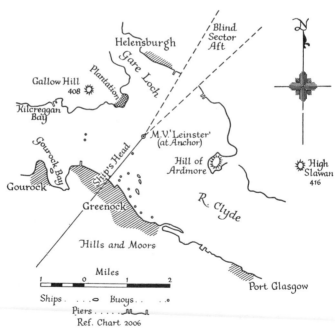

Plate 14. Shielding. *Hill of Ardmore (in the Clyde) appears as an island. Note blind sector (Chapter* 7).

INTERPRETATION OF THE DISPLAY 105

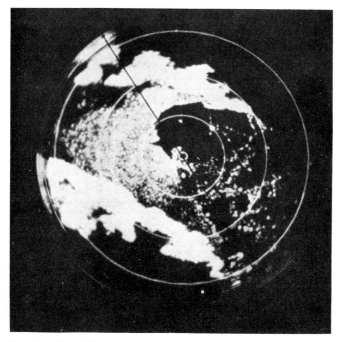

Plate 15. Land and ice. *The PPI of an ice-breaker in the Baltic; a ship, being convoyed, shows south of the centre. Pack-ice frozen solid shows to the NW between two stretches of land; to the SE there is open water with drift floes.*

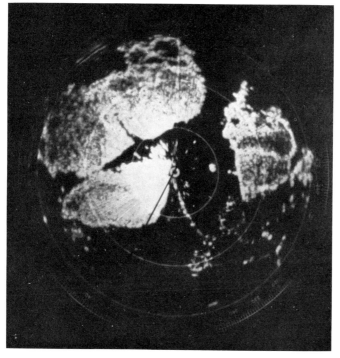

Plate 16. Ice echoes. *To the NW and N is pack-ice frozen solid; floes frozen solid shows to the SE. No echoes are given by fixed sheet ice to the S and SE. Note the ice-breaker's track through the sheet ice.*

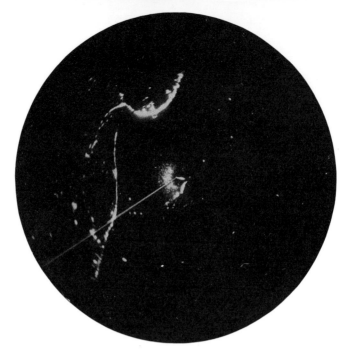

Plate 17. Various types of echo. *The Goodwin sands lie to the south and south-east; Ramsgate cliffs to the north and the low-lying foreshore near Deal to the westward. The performance monitor plume is showing (in an easterly direction), and sea clutter is visible.*

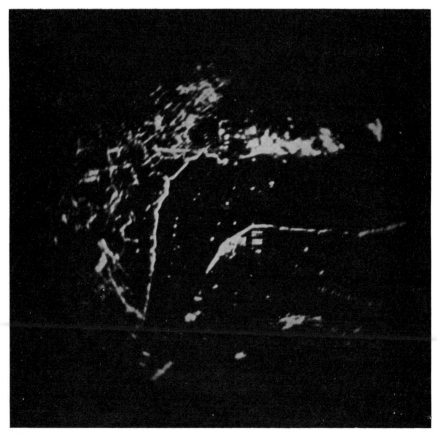

Plate 18. Built-up area. *A reach of the Thames; the huts on the low-lying south bank show a distinctive pattern.*

necessarily be that of its highest point; an escarpment well below the peak may easily return a much stronger echo than a top which is smooth and sloping.

As explained in Chapter 4, it may be taken that surfaces which are relatively smooth will give strong echoes when they are more or less perpendicular to the radar beam, in both horizontal and vertical planes, but poor and sometimes negligible echoes when they are aslant it. In contrast, rough surfaces tend to reflect in a diffuse manner, so that the echoes will be moderate in strength but less dependent upon the angle of view. Thus vegetation (especially in the form of trees) covering a rock face would be likely to weaken echoes when square-on, but when inclined to the radar beam to give stronger echoes than would be obtained from the bare rock.

Whenever there is a land/water boundary, echoes may be expected. Sandbanks and reefs awash may be clearly distinguished, even though several miles distant. The sandbanks in the Thames Estuary and the well-known Goodwin Sands off the Kentish Coast, for instance, give prominent echoes at certain states of the tide (Plate 17). Even in a river, where the lapping of the water against a gently sloping mudbank may not be visible by eye from the ship, it is likely that the waterline will be clearly shown on the radar display. It is hardly necessary to point out here that if the drop in level is sufficient to expose large areas of flats, the radar picture at low water may be very different from that at high water.

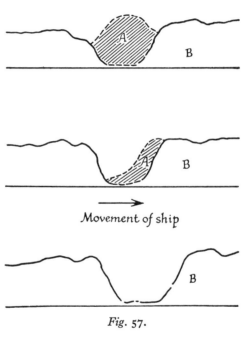

Fig. 57.

From the foregoing description it will be evident that as the ship approaches land, and the echoes from the land extend, the manner in which they do so is dependent on local topography and it is not easy to make more than a rough forecast of the radar picture of a particular area. For recognition purposes no prediction of the radar picture from a chart can be expected to be as effective as previous knowledge (recorded perhaps by a series of PPI photographs taken from different ranges and bearings).

Echoes from ice

The detection of ice is of great importance to ships operating in the colder parts of the world. Ice appears in various forms and, since these produce quite different types of echo on the radar screen, they are worthwhile considering individually.

First, there should be no difficulty about the recognition of smooth flat ice; because of the glancing incidence of the radar beam, this ice returns so little energy to the radar receiver that the corresponding area of the screen will be free from echoes and consequently blank. Only in weather conditions in which the sea is flat calm is there likely to be any doubt that the lack of echo is due to the presence of ice, unless it is well outside sea-clutter range. From seaward, the edge of the ice should give an echo. If such ice is broken up, the floes formed may be expected to have sheer edges giving rise to strong reflections and, because of the floe movement, to appear rather like sea clutter (Plate 15). Ice walls and overhanging edges of fixed ice will also give strong echoes. Should the broken ice freeze together again, the floes overlapping one another, there will be a mass of echoes received from the numerous edges and corners formed where they join. Because of the multiplicity of these echoes, they will produce an effect upon the screen similar to heavy sea clutter. They can of course be distinguished from sea clutter because the echoes will repeat in the same positions from revolution to revolution of the aerial (Plate 16).

Pack-ice frozen into an otherwise smooth icefield will give echoes which will stand out clearly, while the piled-up ice marking the track of a ship which has forced the ice may be detected long after the path has frozen over, even though it is covered with snow. If a smooth icefield is interspersed with patches of water, these may be revealed not only by the edges of ice but by clutter on the water, if there is wind enough to create it.

As with land and sea echoes, judicious use of the gain control is likely to be of great assistance; it will enable the user to obtain an extremely good idea of ice conditions extending over a large area.

Icebergs are often not detected at ranges as long as might be expected. Cases are often reported of bergs looming so large to the eye that it would seem almost impossible for them to escape detection by radar, and yet producing no echo until they are even closer. As mentioned in Chapter 4, this is likely to be accounted for by the berg having smooth sloping sides and possibly also because of sub-refraction which often occurs in regions where ice is prevalent. Sometimes, after having once been detected on the radar, a berg echo may disappear although the range has not increased; this is probably caused by change of aspect, because of either movement of the ship or rotation of the berg. Antarctic bergs have sides which are noticeably sheer compared with Arctic bergs and suffer much less from these effects. The strength of reflection of radar waves from ice on the Grand Banks is 60 times less than from a ship of equivalent echoing area.

Perhaps the detection of ice is most uncertain when it is encountered in the form of growlers. These small icebergs, which show a few feet above the water, are seldom detected at more than 4 miles and return echoes which are almost impossible to distinguish amidst strong sea clutter, due, possibly, to the fact that they are themselves moved by the sea. In general their shape is unfavourable and this combined with their small surface size makes them bad targets. Waves over 4 feet high might obscure a dangerous growler despite expert use of sea-clutter control and differentiator. As a warning of such ice in the vicinity of a ship, a radar set is unreliable, and as the bulk below water of some growlers is quite sufficient to damage a ship, every precaution is necessary when navigating in ice conditions.

Echoes from the sea

Sea clutter is generally regarded as a nuisance, and as such is dealt with under Prejudicial Echoes in the next chapter. Sometimes, however, clutter can give useful information; for instance, it can show tide rips and lines of demarcation between currents, possibly more distinctly than they can be seen by eye. Sometimes near a coast the echo from a tide rip is so strong and of such a length that it may prove alarming if detected in poor visibility, since it could be mistaken for the coast itself. In some rivers it is possible to distinguish the down-going from the up-going current, by the different textures of the clutter. The presence of overfalls, the disturbed water which occurs above a sudden rise in the sea bed, can often be detected, and in fact in shallow water near a coast some navigators claim to be able to follow the lie of the sandbanks, although under water, by the clutter above them. In deep water the distance between wave crests is sometimes sufficient for the wavelength to be measured by radar. This may be of use in connection with meteorological reports.

Echoes from clouds

These also are dealt with in the next chapter as unwanted echoes. They may, however, give useful information since detection on marine radar sets of pronounced weather fronts and troughs of low pressure is on record.

ARTIFICIAL TARGETS ON LAND

There are many targets on the land to be considered other than those provided by nature. Large artificial structures, such as breakwaters, offer a most distinctive key to the recognition of a coast, since their regularity of design contrasts so strongly with the average coastline. Furthermore, for the reasons that have been explained in Chapter 4, buildings often return stronger echoes than land masses many times their size (Plate 18). Hence when echoes at long range are obtained, it does not follow that they come from high land or from cliffs; they may be due to groups of buildings. A good example of this is Flamborough Head on the Yorkshire coast, where the cliffs rise to 200 ft. and where approaching from the south or east, the first echoes might be expected to be those from the cliffs. They are more likely, however, to be echoes from the hillside town and holiday resort of Bridlington, a few miles to the south-west, which can be detected at a range of over 20 miles.

Buildings, roads, &c.

At fairly close range, the details of a town are often revealed on a radar picture, though this picture is likely to differ from a visual plan in three main respects: first in that only the nearer sides of buildings will be shown, the other parts being in shadow; secondly, because buildings in the foreground will shield those behind unless the latter are higher; and thirdly, because of reflections between the buildings themselves, though this effect is a minor one. Any group of buildings, regularly spaced and standing by themselves, will form a feature that is conspicuous because of its regularity of pattern. It should be possible to trace wide streets and, away from towns, the courses of roads and railway lines, not so much because of echoes they themselves may return, as because of those returned by adjacent walls, metal fences or lighting standards. Bridges across a river give pronounced echoes (Plate 20). Hedges and lines of trees will give echoes which are clearly defined at short range, and marsh land with reeds will return echoes over its whole area.

Steel structures

A large cylindrical mass, such as a gasholder, at close range may produce an echo in the shape of a crescent (it is of interest to note that when using a pulse-length of 0·1 microseconds the framework of a collapsed gasholder has been observed as a ring). Fairly strong echoes may be expected from aerial towers and masts and other erections composed of steel girders. Dockside cranes and metal lamp standards are conspicuous. A pier extending out to sea will when seen from the side give a good line echo, but when viewed end-on may appear as a series of dots; this will probably be due to shielding of the bulk of the pier by the part further out to sea, leaving the lamp standards along it to be separately depicted.

Chimneys and towers

Tall factory chimneys which are not tapered, generally give good echoes at medium range. When they are not surrounded by buildings giving strong echoes it is sometimes possible to sort them out from other echoes by temporarily reducing the gain, and thus they may be used as a landmark for navigation. The effect of reducing gain is illustrated in Plate 19. Echoes from most lighthouses have been found, however, to give a disappointing range, disproportionate to their height. As stated in Chapter 4, this is probably due to their tapered (or conical) shape.

Waterfronts

Docks and dock entrances are clearly shown if not too distant; the limitation of discrimination, as described below for the case of ships, applies equally here. In rivers, it is often impossible to detect objects beyond bridges, which give strong echoes themselves, and cast pronounced radar shadows. Ships which have low aerials may, however, see through.

Location of the arches of a bridge may be possible if the piers on which the pillars rest project sufficiently.

Thus, where a position cannot be sufficiently well established by the natural features of an area, recourse may be had to other landmarks. For off-shore and river pilotage, familiarity with the radar appearance of the locality will be of the greatest value in establishing and checking the ship's position.

Cable crossings

Overhead cables above a river give a small and not very strong echo from the point on the cable at which the radar beam is at right angles to the run. If the ship's course is at right angles to the cable run the echo will be fixed. If not, however, the echo will move out from the bank on a steady bearing and may be misleading if unexpected. The same effect may be caused by a bridge which is not at right angles to the channel.

ECHOES FROM SMALL ISOLATED TARGETS

Targets which are large in comparison with the horizontal beam-width of the aerial, e.g. a coastline or a large island, will give an echo the shape of which to a considerable extent will resemble the contour of the original. The shape of the echoes of small targets will be determined solely by the characteristics of the radar set. This will apply to the echoes of buoys and small boats at any range and to ship echoes except when the targets are large and very close. Other targets comparable in size with ships will record similar echoes, whether they are rocks, icebergs, houses, chimneys or lighthouses, as long as they are isolated. The reason for this is easily explained and it is important that it should be understood. The same factors cause the shape and size of the echo of a small target to change as the range alters, so this knowledge will assist interpretation and identification of echoes.

As mentioned in Chapter 1 the three factors which affect the size and shape of the echo of a small target are spot size, horizontal beam-width and pulse-length. When the combined effects of these three factors are considered it will be found that, on first detection, the echo of a small target will be governed by beam-width and spot size; it will, therefore, be elongated in azimuth. At shorter ranges, the minimum radial width will be governed by pulse-length rather than spot size, and the elongation in azimuth, while having the same angular extent, will be reduced in linear width. This will suggest that a small echo is shaped like a box bounded by two radii and two arcs. This, though not an accurate conception, is a useful approximation. It will be remembered, however, that, since the 'box' has to be drawn by the somewhat blunt 'pencil' of the spot of light, its dimensions in all directions will be increased by half the spot width. Its corners will be rounded partly for the same reason.

Since this box governs the shape and size of the smallest echo, it will also decide whether the echoes from two targets close together will be displayed separately or as a single echo. In fact, it governs the discriminating ability of the equipment as well as its power of indicating the shape of a target, i.e. its resolution. For this reason the box may be called the *minimum area of resolution*.

One further point must be made. In the above explanation only very small targets, such as buoys, have been considered. With a somewhat larger object, such as a ship, the same general conditions will apply, but the echo may be slightly larger. If the example of a fair-sized ship with an aspect of 45° is taken and its two ends are considered separately, it will be seen that 'boxes' will be formed by the echoes from both the bow and the stern. Of course, they will also be formed by echoes from the midship position and all the boxes will join up to form an echo larger than that of a buoy. The resolution will not be good enough, however, to make it clear that the stern is further away than the bow. This will only occur when the ship is very close and the shortest range scale is used so that the minimum area of resolution is small compared with the projected area of the target. It will now be realized that this may be achieved either when the target is very large or when the beam-width, pulse-length and spot size are all very much reduced.

Echoes from ships

The behaviour of a ship echo, so far as its shape and size are concerned, place it in the category of 'small isolated objects' just described.

The *strength* of the echo reflected from a ship will depend on many factors, but it is possible to make rough divisions between classes of ships in terms of their echo strengths. On the more elaborate radars used in the armed services, the strength of an echo can be measured directly, and as its range is known a fair guess can be made from a series of graphs as to the class of ship represented by an echo. But on merchant navy radars, the most satisfactory criterion of signal strength is the range of detection, and this offers a rough guide to the size of ship whose echo is represented. Figures such as 5 miles for a drifter, 10 miles for a 1600-ton collier, and 18 miles for one of the biggest passenger liners, might be regarded as typical. Such ranges are of course subject to such variables as the aspect of the ship, the depth to which it is loaded and the propagation conditions at the time. In conditions of super-refraction, many cases of ships being detected at 25–30 miles have been reported.

From the description of the change of shape of echoes with range, it is apparent that at medium range the echo of a ship may, irrespective of her true heading, give the impression that she is beam-on (Plate 21). This impression should be guarded against, and her course (if she is under way) be determined by one of the recognized plotting methods which are described in Chapter 10. Closer in, the converse may apply and the ship appear to be end-on. On a large scale, with the ship close, it is possible that the near side of her hull will be outlined (Plate 22).

Echoes from buoys and other small objects

The echo of a small object, such as a buoy, changes in brightness from the time it is first within range of the radar set. When first detected the echo will paint only occasionally as the aerial rotates. At this detection range several causes will contribute to preventing the echo from being seen regularly. For a floating object such as a buoy, there are continual changes in aspect due to its movement by wind and waves, while the waves may actually screen it from the radar beam on occasion. Reflections of the beam from the sea, by interfering with the direct beam, may also be a cause of fluctuations (see page 69).

Yet another cause is local variation in the refractive index of the atmosphere, causing scintillations in the echo. Were the aerial to be stopped on the bearing of the buoy, and the echo observed on an A-scan (on which echo heights are proportional to echo strengths: see Appendix IV) it would be seen that the height of the buoy echo is subject to continual and irregular variation. Now, as the distance closes the average strength of the returned echo will increase, and it will thus commence to paint regularly and more brightly as the aerial rotates.

With the buoy still closer, a stage will be reached at which the brightness of all paints will be at the limiting level or saturation. Such a consistent echo is said to be *firm* or *solid*. If the echo is followed right in (assuming little or no sea clutter is present) it will eventually vanish into the central spot, at the minimum range of the set.

Some buoys are fitted with radar reflectors to increase their range. It should be mentioned here that it is a characteristic of the radar reflector that, when the sea is calm or nearly so, the echo of a radar buoy will fade and may even disappear at certain ranges, to re-appear when the range is altered. This phenomenon is fully described in Chapters 4 and 12. Some typical maximum ranges for various classes of buoy are given in Appendix I.

As regards echoes from other small objects, wooden boats of some size give echoes comparable with those of buoys, but may be distinguished from them if they happen to be moving. Low-flying aircraft and occasionally flocks of birds can be detected and, if the water is dead calm, echoes from seagulls sitting on the water, or from fish breaking surface, can be seen at a few hundred yards.

Distinguishing between targets

The ability to distinguish between the echoes of land and ships and between ships and buoys and so on will obviously be of great importance to the navigator. Distinguishing between land echoes and those of small isolated objects is usually straightforward. At anything but very short range, as has been said, the latter are fairly sharply defined by the minimum area of resolution. Land echoes, on the other hand, even when just perceptible, usually possess an individual shape, which alters considerably with change of range or aspect, and they are rather more diffuse. If own ship's position is known there will seldom be any difficulty, and even if it is not, the distinction should be easily drawn.

A distinction between kinds of target in the 'small and isolated' category cannot be made on a basis of the shape and size of the echo, except occasionally in the case of radar reflector buoys. It will, however, often be possible to base it on what may be called operational evidence. This may be made clear by considering the case of a radar-fitted ship whose position is known and which is proceeding along a coast.

Two echoes are observed to make their first appearance on the screen, the one at a range of 2 miles, the other at 9 miles. The radar observer, knowing the navigational marks in the vicinity, has little hesitation in identifying the former as being almost certainly that of a particular buoy or of a small boat, and the latter as that of a ship. After a little while he is able to confirm his first opinion, and his belief that the nearer echo is from the buoy. The processes of thought which lead him to those conclusions are the following:

> First, there is circumstantial evidence. A buoy was known to be in the approximate position in which the nearer echo was seen, and its position was checked on the chart. On the other hand, no buoy or other fixed object was expected in the position of the further echo.
> Secondly, there is the significance of detection range. The observer knew that 2 miles was a reasonable detection range for a buoy of the type marked on the chart. He also knew that while a ship of any reasonable size could be detected at 9 miles, no ordinary buoy or small boat would appear at this range.
> Thirdly, the echoes were plotted for movement. This showed that the near echo was stationary, while the further one was moving at a likely speed for a ship.

These methods are applied, perhaps subconsciously, by all competent observers. It will be noticed that conclusions are drawn as much by rejection of improbable hypotheses as by any positive information; in the above case, the evidence establishes, not so much that the echoes must be those of a buoy and a ship, as that they are unlikely to have been produced by other objects. The desirability of an observer knowing his average detection ranges is obvious. At the same time, the observer should never over-estimate the reliability of his deductions.

Where two or more echoes are at the same range, it may often be useful to distinguish between the weaker and stronger echoes by gradually turning down the gain control. For example, if three stationary echoes are seen, which are equal in size and brightness but are believed to be from two ships anchored near a buoy, the weaker buoy echo will disappear first as gain is reduced. The same method will serve to distinguish between two buoys, one of which is fitted with a radar reflector. The fading characteristic already mentioned may on occasions assist in the identification of a radar reflector buoy. It is hardly necessary to add that when gain has been

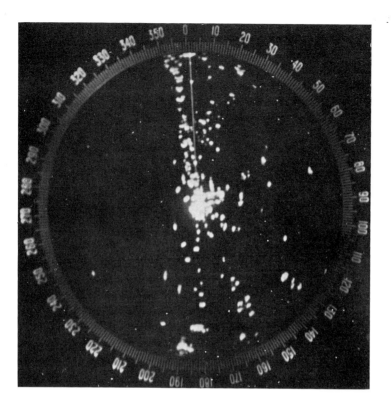
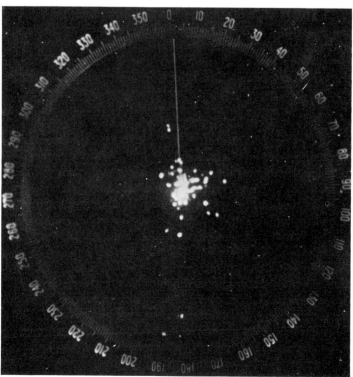

Plate 19. *Clearing the picture by reducing gain; 1-mile range scale.*

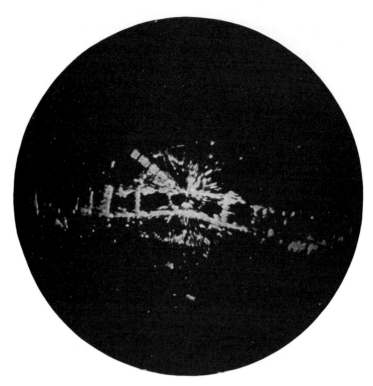

Plate 20. Bridges in the Thames. *The buttresses can be seen in some cases; side-lobe and indirect echoes are present. Note side-lobe echoes and monitor plume (paddle stopped—see p. 245.)*

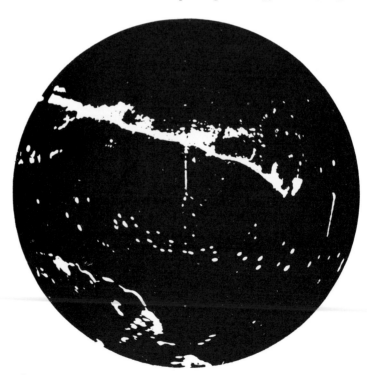

Plate 21. Ships at medium range. *Ships anchored in the Thames, showing beam-width distortion; the radar set is situated at the end of Southend Pier.*

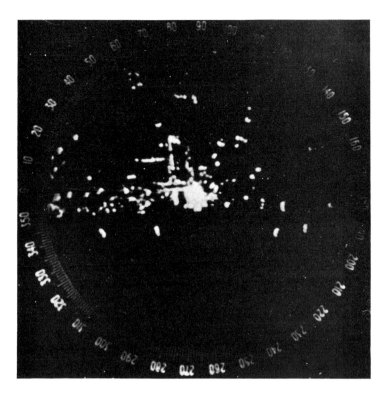
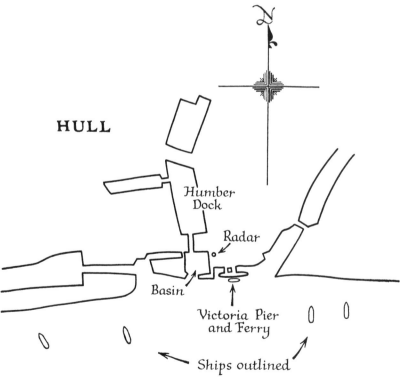

Plate 22. Ships at close range. *A shore radar PPI on $\frac{1}{2}$-mile range scale. The ship echoes indicate the directions in which they are lying.*

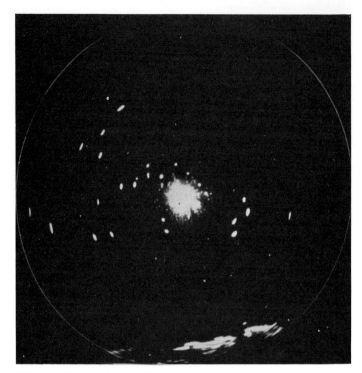

Plate 23. Buoy echoes. *The majority of the echoes are from buoys, those with radar reflectors having stronger echoes.*

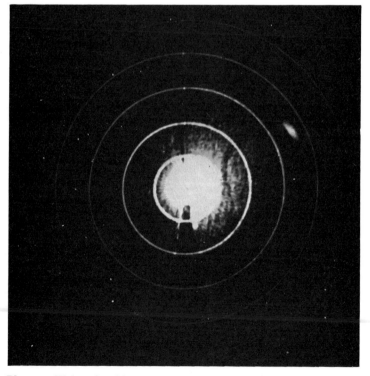

Plate 24. Wake of a ship depicted by the smoothing of clutter. *The ship to the north-east is just completing a turn of* 180°.

reduced for the purpose of distinguishing between echoes, it should be restored to its normal setting immediately the check has been made.

No simple method has yet been devised for differentiating between two stationary echoes of comparable strength at the same range: for example, a buoy and a moored wooden boat.

Plate 23 shows typical echoes of ships and radar reflector buoys.

Appreciation of movement

Movement of a ship cannot be appreciated nearly so readily from the radar screen, as it can be by direct observation. When a ship is seen by eye, a number of tell-tale pointers indicate movement: the bow-wave, the wake, change of position relative to a nearby coastal feature and so on. On the radar screen, not only are these pointers absent or much minimized in effect, but the angular movement of the echo relative to the eye is much reduced when the echo is near the centre of the screen. This effect is particularly noticeable for targets close to the ship; the eye may be immediately attracted by a small boat moving past some buoys, yet on the radar screen, in spite of the presence of the buoy echoes as markers, some seconds may elapse before the moving target is picked out.

Echo trails

Sometimes if the ship echo is strong and moving fast, its movement may be very apparent from its afterglow or 'trail,' the whole giving an effect suggestive of a tadpole. It should, however, be noted that, on the Relative display, this indicates relative movement, and that the movement of the radar-fitted ship herself may give a trail to the echo of a stationary target.

From what has already been said regarding ship echoes, it will be understood that one of the difficulties confronting an observer using radar as an aid to avoiding collision is its inability to show the aspect of ships from the shape of their echoes. Another factor on the Relative display is that the rate and direction of the movement of an echo on it may be very different from the true movement of the ship causing it. In that display, own ship remains at the centre of the PPI; the sea and the land and other fixed objects pass by at her speed; the echoes of moving objects cross the screen with a movement which is the vector difference of their course and speed and that of own ship.

In the True motion display, own ship and all other moving objects move across the screen with their proper motion, while fixed objects remain stationary or nearly so. This highlights the two main objects of the True motion display, which are to enable fixed objects to be distinguished at a glance and to have echo trails which show the true courses of ships instead of the relative.

Whether the echo trails will be visible or not will depend upon the persistence of the PPI afterglow and the rate at which the echo moves across the screen. Obviously, the latter depends upon the speed of the other vessel and the scale in use on the PPI. Trails will be difficult to see at ranges in excess of 6 miles and generally true motion facilities are not provided on range scales in excess of 12 miles.

If sea clutter is present, the wake of a ship may be manifested by a smoothing of the clutter, as in Plate 24.

Methods of measuring movement of ship echoes accurately are dealt with in the chapter on Collision Avoidance.

The value of experience

This chapter has attempted to summarize the characteristics of those echoes which, if recognized, can be of navigational assistance to the radar user. The variety of other effects, which may appear on the PPI but will seldom be of use, are described in the next chapter. From the story of the useful echoes alone, however, it will be appreciated that skill and experience are prime requisites of an observer; the attainment of these rest with him. The art of radar interpretation is still young and this record makes no claim to be complete. In the course of their duties, radar observers may make discoveries which will add to the breadth and importance of the art.

7

UNWANTED ECHOES AND EFFECTS

IN certain circumstances false echoes may appear on the PPI and the clarity of the display may be interfered with by undesired effects. These echoes can be distinguished from the useful echoes described in the last chapter, and in many cases they can be controlled. Their causes can be grouped into those which arise from sources outside the ship and those which are due to local sources, such as the ship itself or the equipment. It is convenient to examine them in these categories and this chapter will attempt to show how each can be recognized, and where possible dealt with.

SEA AND WEATHER EFFECTS

Sea clutter

When sea clutter is severe it can be most prejudicial to the detection of wanted echoes within its area. When the strengths of echoes from small objects such as buoys are no more than comparable with those of reflections from the waves it may not be possible to distinguish them. The use of the gain and sea-clutter controls in this connection has been dealt with in Chapter 3.

The appearance of sea reflections is unmistakable. They usually cover a considerable area which contains a very large number of small echoes each of which changes in position and brightness from one revolution of the scanner to the next. In almost all sea conditions there will be an area of sea clutter around the ship. The fronts of waves give stronger echoes than the backs and therefore the clutter area will not be symmetrical about the ship, but is likely to be roughly oval (Plate 25) the greater part of it lying to windward. The extent of the area will vary according to the state of the sea and may extend to 3 or 4 miles from the ship when the sea is rough. A choppy sea produces closely spaced returns which may be so dense near the centre of the screen as to give the effect of an almost solid echo. If the movement of the sea is largely due to swell, there will be a wider space between the echoes although they will extend to some distance. Lines of rollers sometimes produce very distinct echoes (Plate 26).

The use of the sea-clutter control has already been described in Chapter 3, but it should be noted that this control operates equally in all directions and therefore takes no account of the fact that the sea-clutter area is unsymmetrical. When it has been adjusted to the best setting for any particular circumstances it is probable that a crescent of clutter will remain at the edge of the area on the windward side.

Narrow aerial beam width and short pulse length act favourably on sea clutter, as does also the logarithmic receiver. Sea clutter increases with aerial height, but this is usually accepted in the interest of a clear all-round view from the scanner.

As was mentioned in Chapter 6, reflections from patches of disturbed water remote from the ship, are sometimes observed. While it is unlikely that clutter at this range will be strong enough to obscure echoes of navigational importance, this possibility should not be excluded; thus, echoes of boats fishing over a shoal might be obscured by local sea clutter.

In certain extreme conditions of super-refraction, the radar beam may bring in sea-clutter echoes from distances outside the normal range of the radar (multiple-trace).

Rain and rain clouds (see also Chapter 5)

The water content of precipitation, whether it is in the form of rain, hail or snow, is a conductor of electricity and hence is almost opaque to radio waves. As with metal structures, the effect on the radar beam will be reflection and scattering and there is almost always enough of the former to give echoes on the PPI. The ship herself may or may not be within the precipitation area and the first type of precipitation to be considered is *falling rain* at a distance from the ship. This can usually be recognized without difficulty. Rain squalls of the type encountered around the British Isles and in other temperate areas usually display filmy soft-edged echoes, often likened to cotton wool. The effect is well illustrated in Plate 27. Heavy rainstorms will give echoes of a more solid nature (Plate 28), while tropical storms will produce echoes with sharply defined edges, the harder edges being those nearest the ship.

The effect of rain echoes varies considerably, but it will seldom be severe enough to obscure ship echoes if proper use is made of gain control. The strength of *rain clutter* depends on the intensity of the storm, that is upon the size of the drops and the spacing between them. The drops scatter the outgoing and returning radio waves, which lose more energy the further they have to penetrate through rain. The strength of the echoes of targets inside the squall will therefore be progressively diminished and the reduction of energy suffered by the wave will naturally reduce the strength of the echoes of targets beyond the rain. Although, with the gain at its normal setting, the rain echoes may be strong enough to obscure the echo of a ship inside the squall, a reduction in gain will usually permit the strong and solid ship echo to be distinguished. With intense tropical rainstorms this distinction will be more difficult but, as has been said, it is seldom impossible. Detection of targets beyond the rain may, of course, require slightly higher gain than normal, since their echoes have been weakened but not obscured.

Intense storms may be observed out to the full range of the set and they may be sufficient to prevent the echoes from land beyond them being received except through gaps in the rain. The echoes they display may be quite similar in appearance to those of land. When the area of a heavy rainstorm is very limited its echo presents an appearance very similar to that of a ship, though usually the experienced observer will detect alterations in shape and consistency which will enable the distinction to be made. Water spouts have a similar radar appearance but are smaller.

Reduction of rain clutter may be achieved by a circularly polarized aerial and by a logarithmic amplifier, particularly when followed by a differentiator. (See Chapter 2 and Plate 4.)

Rainstorms of all kinds are usually moving and their tracks can be plotted. The direction of the wind, and sometimes the speed at which they travel, may help in their identification. Their echoes can be seen travelling over the echoes of land and ship targets.

When the radar-fitted ship herself is within a precipitation area, the rain will reduce detection ranges of all targets to an extent which of course depends upon its intensity. It may add to the effect of sea clutter and so make the detection of nearby echoes more difficult. As in all cases where the target is inside the precipitation area, the deterioration of the picture may be reduced by careful manipulation of the gain, sea- and rain-clutter controls. (See Plate 4.)

Even if no rain is falling, rain-filled clouds will give strong echoes. These may be strong enough to obscure other echoes with which they coincide on the PPI; here again it is a question of the relative strength of the echoes although the cloud may be above the ship. Reduction of gain may disclose the ship or land echo. The radar appearance of low clouds is often similar to that of a coastline.

Hail

The appearance on the display of hailstorms is very similar to that of rain. The strength of the echoes depends upon the size of the hailstones. For small hailstones, echoes will be weaker than for a corresponding rainfall, i.e. than if all the hailstones were converted into water drops. Hailstones larger than about $\frac{1}{4}$ inch in diameter will give echoes exceeding that of corresponding rainfall, but such conditions will be unusual. The echo strength of other targets is affected much less by hail than by rain of equal intensity.

Snow

In general, echoes from snow are not likely to be troublesome, except in very heavy precipitation in cold climates where snow falls in single crystals. In temperate climates, where the snow crystals are aggregated into large flakes, intense snowstorms can give much stronger echoes than would a corresponding rainfall; but since in practice the corresponding rainfall intensity is likely to be low, it may be said that snow echoes will be of less consequence than those from rain. The echo strength of targets is unlikely to be affected noticeably by snow.

Sandstorms

In such areas as the Red Sea and the Persian Gulf, visibility may be much reduced by sandstorms. Here radar is of great value because the effect of such sandstorms on detection ranges is barely appreciable. In other areas such as the west coast of Africa, where sandstorms may blot out the shore (Harmattan haze), a similar slight reduction in radar detection ranges is observed.

Fog

In warm climates fog will have little noticeable effect on normal detection ranges, unless it is so thick that visibility is practically nil. In colder climates, however, dense fog will have an appreciable effect; for example, it has been calculated that in a visibility of 100 ft. (300 metres) a target normally detectable at 27 n.m. (50 kilometres) would have its detection range reduced to 16 n.m. (30 kilometres).

Echoes from fog are negligible. Observers have claimed, however, to be able to detect the presence of fog banks from faint indications on the PPI.

Atmospheric discontinuities

This is the general term used to describe irregularities in the lapse rate or rate of change of temperature or humidity in the atmosphere. From Chapter 5, where such matters are dealt with, the impression might be received that these intangible effects only cause bending of the radar beam, while atmosphere echoes are due only to precipitation in one form or another. However reports have been received of radar echoes of an indefinite character, rather like second-trace echoes of land, but extending over several miles, when there has been no possibility of precipitation or of second-, third- or later trace echoes. It is believed that they may be due to sharp changes in the lapse rate of the humidity, probably at a few hundred feet above sea-level. The echoes are usually seen at ranges of over 10 miles and may be 20 miles or more in length; sometimes they take the form of concentric rings of 6 to 8 miles diameter.

Non-standard (anomalous) propagation

Anoprop and the meteorological conditions which are favourable to it have already been described in Chapter 5. It is most commonly met with in the form of *super-refraction* when objects may be observed at unexpectedly long ranges. This, of course, will often be of great advantage. Some slight disadvantages are also involved; for example, detection range will no longer be a guide to the identification of a target.

It is important to remember that these long ranges are solely the result of particular climatic conditions and cannot necessarily be repeated. They should not be confused with *average detection ranges* which may be used as an indication of full efficiency on the part of the set. In certain areas and at certain times of the year, however, super-refraction conditions are met with more often than not. If the average detection ranges of a variety of targets are known they will be an excellent guide in deciding whether super-refraction is present or not. The difference is generally so marked that there is no room for doubt. Knowledge of the area and the current meteorological conditions will be additional guides.

In super-refraction conditions, it is quite possible for ships, coastlines, &c., well below the radar horizon for standard conditions to be exhibited out to the full range of 30 miles or more of the set. Echoes may be received from longer ranges still, but if so will be displayed as *second-trace echoes*, which are described below.

The effect of *sub-refraction* on the radar beam is usually to reduce detection ranges considerably. It has been less frequently observed than super-refraction and occurs mainly in cold areas, particularly icefields. Its presence may be suspected when poor results are being obtained, despite local indications of good set performance, and when the meteorological conditions are appropriate. An extra degree of caution should then be exercised.

Second-trace echoes

It will be obvious that on many occasions echoes will be sent back from targets just outside the maximum displayed range of the radar set. These are not registered on the PPI because, by the time they arrive, the spot has returned to the centre of the display and is awaiting the transmission of the next pulse. If, however, a target returning an echo is sufficiently remote, the echo may not arrive until after the next pulse has started. If this occurs and the echo is strong enough it will be painted on the PPI on its correct bearing but, obviously, not at a point which represents the actual range of the target. Whether or not the echo will arrive while the trace is moving out will therefore depend upon the interval between the pulses. The following table shows the target range corresponding to various inter-pulse intervals.

TABLE VI

Repetition frequency (pulse/sec)	Inter-pulse intervals (seconds)	Equivalent target ranges (n.m.)
500	1/500	162
1000	1/1000	81
1500	1/1500	54
2000	1/2000	40

This means that on a set which has a repetition frequency of 500 pulse/sec and a maximum range of 30 miles, echoes from targets between 30 and 162 miles distant cannot paint. If the repetition frequency were 2000 pulse/sec this restriction would only apply to targets between 30 and 40 miles distant. In this latter case an echo from a target of 40 miles range would appear at the centre of the tube because the (second) trace would at that moment be about to start on its next travel. A target at 50 miles range would produce an echo when the (second) trace had moved to 10 miles. This explains why these echoes are known as *second-trace echoes*. It will be observed that the probability of receiving second-trace echoes increases as the repetition frequency increases. With present-day marine radar equipments it will be unusual under normal meteorological conditions for an echo returning from a target at a range of more than 50 miles

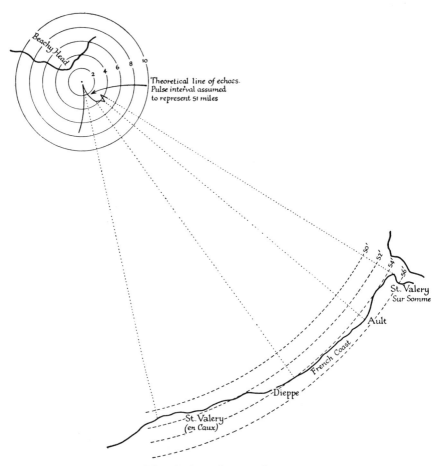

Fig. 58. *Second-trace echoes.*

to be strong enough to produce a paint on the PPI. As none of these sets has a repetition frequency (on its long-range setting) greater than 2000 pulse/sec it is unusual for second-trace echoes to be received; if, however, super-refraction conditions are present it will be quite possible, and in fact several such occurrences have been reported. An actual example of such an occurrence is illustrated in Fig. 58. Here a radar-fitted ship off Beachy Head on the south coast of England, received second-trace echoes from the French coast which took the peculiar form shown. The radar set in question had a nominal repetition frequency of 1500 pulse/sec.

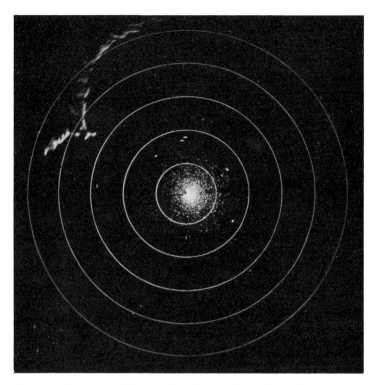

Plate 25. Sea clutter. *The majority of clutter is to windward; range rings 2 miles apart.*

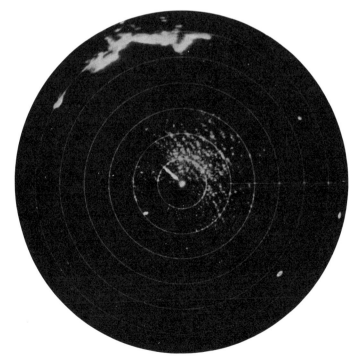

Plate 26. Lines of rollers.

Plate 27. Light to moderate rain squalls.

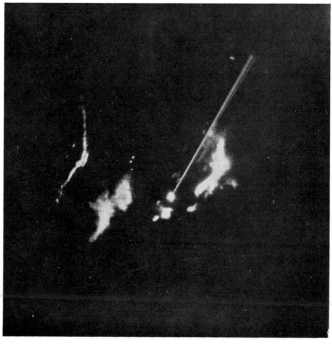

Plate 28. Heavy rain squalls. *Taken off Lowestoft.*

Plate 29. Multiple echoes. *A 700-ton ship at 200 yards range; the echoes to the right are ducks taking off.*

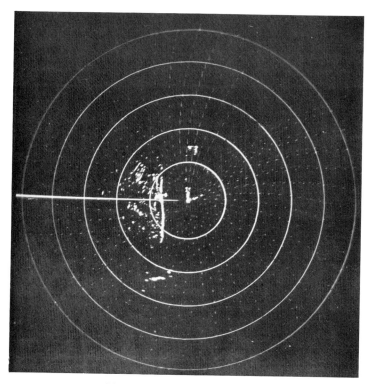

Plate 30. Radar interference.

The distortion shown can be explained very simply by considering a coastline that is absolutely straight and on which the nearest point to the ship is 64 miles distant. If there are two other points on the coast, one on either side of the first, and both distant 74 miles from the ship, then the radar set will record the range of the centre point as 10 miles, and of the other two as 20 miles. A simple drawing will show that this straight coastline will appear on the PPI as a pronounced curve towards the observer (Fig. 59). Similarly, a coastline curving away from the observer may

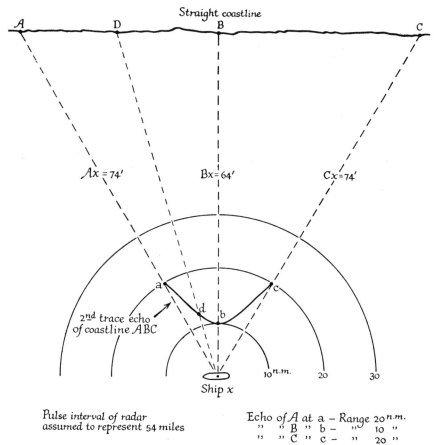

Fig. 59. *Distortion in second-trace echoes.*

appear as a straight line. The general tendency, therefore, of the trend of a coastline as it will appear in a second-trace echo is that it will be bowed towards the observer, headlands being accentuated and indentations smoothed out or even reversed in their trend.

The second-trace echo of a ship steering along the line ABC in Fig. 59, on a parallel or opposite course from ship X, would describe the curve *a b c*. Successive radar bearings and ranges of second-trace echoes, plotted on a relative plot (see Chapter 9) would give a relative track of similar shape to that curve. This shape and sometimes the speed of the other ship obtained from the plot will be an indication that the echoes are in fact second trace. The relative track of a ship pursuing a steady course will be curved on the plot except when her bearing is steady.

As has been said, these effects are unusual; nevertheless it may be well to note that third-, fourth- and even fifth-trace echoes have been recorded. It will have been realized that on a set with a 30-mile maximum range and a repetition frequency of 1500 pulse/sec second-trace echoes may be recorded of targets between 54 and 84 miles range. Beyond the latter distance they cannot be recorded until the third trace starts. The following table sums up the ranges of targets which may produce echoes on the second, third, fourth, &c., traces of a set with the characteristics mentioned:

TABLE VII

Trace	Range of target (n.m.)
First	0–30
Second	54–84
Third	108–138
Fourth	162–192

Such echoes may be mystifying if their nature is not recognized. The possibility of putting such echoes to a useful purpose is discussed in Chapter 8. They are unlikely to be seen at all unless a considerable degree of anomalous propagation is present, and if by their appearance they are thought to be ship echoes the best guidance is obtained from plotting their tracks and measuring their speeds. (On some sets, the repetition frequency can be varied; if this is done, no movement of ordinary echoes will be observed, but second-trace echoes will be shifted towards or away from the centre of the screen.)

EFFECTS DUE TO OTHER SHIPS

Multiple echoes

When another ship is passing on an opposite or a parallel course and at close range, say within a mile, a second echo will often be seen beyond the normal echo of the other ship and at double its range. Occasionally there may be a series of such echoes all on the bearing of the other ship and at equal intervals of range (Plate 29). This effect occurs only when the echo from the other ship is strong enough to be reflected back and forth between the hulls of the two ships and, when it returns for the second or third time, it is still strong enough to paint. Owing to the necessity for the reflecting surface to have a high degree of response, multiple echoes are likely to occur only when the ships are almost exactly beam to beam. Their appearance is unmistakable and they are not likely to be more than a very minor distraction to the observer.

Interference from other sets

Transient echoes sometimes appear on the radar screen in the form of curves or spirals of dotted lines which may cross the screen and change in shape from revolution to revolution of the trace. A typical pattern is illustrated in Plate 30. This effect is caused by interference from another radar which is operating on a nearby radio frequency. The indications are so distinctive that they can hardly be mistaken for any other source of trouble, and the explanation of them is straightforward.

Each of the small dots or pips is due to the reception of a pulse from the interfering radar. If the two radar sets had identical pulse repetition frequencies, the interval of time between the start of the trace on the receiving radar and the reception of the interfering pulse would always be the same. The echo of the interfering pulse would, therefore, always appear on the receiving PPI at the same range and a succession of such echoes would form a ring. If, as is more likely, the pulse repetition frequencies are slightly different, successive echoes will appear at regularly increasing or decreasing ranges and a series of them will form a spiral shape instead of a ring. Should the difference between the repetition frequencies be large, the pattern formed will be so confused that the pips will appear to be distributed in random fashion on the screen. The number of pips that can be seen will be very much more when the receiving set is on the long-range scale than when it is on the short and they will, therefore, be much more apparent. On the shorter range scales where the trace is moving faster, the pips will be drawn out into thin lines which are likely to be so faint as to escape detection unless specially looked for. In pilotage waters where ships are likely to be close together and most numerous, it might be expected that interference would be more serious. As, however, ships will probably be using the shortest range scales at that time the interference is unlikely to be disturbing.

The probability of this type of interference occurring is governed by a number of factors of which the most important is probably the width of the band of frequencies which the radar set will receive. The frequencies employed by different ships are purposely spread over the marine radar band, of which any one radar set is able to receive only a portion. In practice interference may perhaps be received from about one ship in twenty. Its absence cannot, therefore, be taken as an indication that a ship whose echo is on the screen has no radar in operation. Modern radars with 'Bright displays' usually include means of suppressing radar interference.

Interference from shore radio installations or a ship's own transmitter has very occasionally been observed, but it is so infrequent as to be of little consequence.

EFFECTS FROM THE SHIP'S STRUCTURE
Shadow areas

Metal objects are opaque to radio waves: the radar beam cannot pass through them. According to their shape and the angle at which the beam strikes them, the radio waves will suffer scattering (dispersion), diffraction, or specular reflection (as a mirror reflects light), but the area directly beyond the obstruction will be in shadow (see Chapter 4). The *horizontal width* of the shadow sector depends largely upon the angle which the obstructing object subtends at the aerial in the horizontal plane, that is to say, upon the width of the obstruction and its distance from the aerial. It

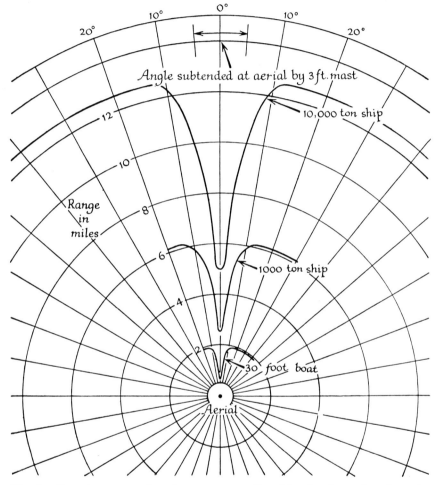

Fig. 60. Detection ranges of various targets as affected by the obstruction of a 3-ft. diameter mast 20 ft. from the aerial.

cannot, however, be treated as a simple problem in geometry because the radar beam does not emanate from a point but from the whole of the width of the reflector, and because the wave cannot be regarded as being confined within absolutely straight lines. The effect is complicated and is still a matter requiring further investigation.

The word *shadow* has been used deliberately to suggest that in the area beyond the obstruction there will be a reduction in the intensity of the beam but not necessarily a complete cut-off. If the angle subtended at the scanner is more than a few degrees, however, there will be a blind sector. The majority of obstructions met with in a ship, masts, samson posts, funnels, &c., affect the whole of the useful *vertical* beam-width of the aerial, but an object such as the cross-trees, limited in vertical dimensions, may not. The aerial may be able to 'look' over or under it, and the shadow

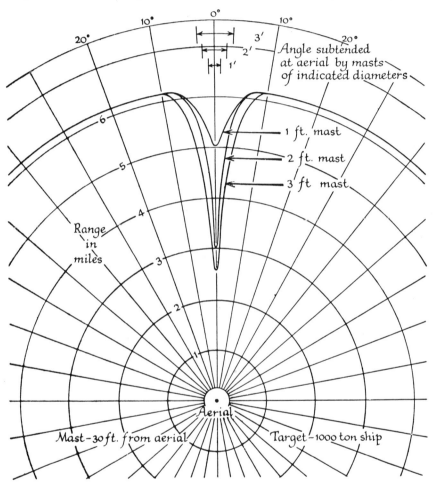

Fig. 61. *Detection ranges as affected by the obstruction of masts of different diameters.*

area will not then extend the whole way from the ship to the horizon. In the vast majority of modern fittings the scanners are placed above every obstruction except perhaps a light topmast. The effect of blind or shadow sectors is so important, however, that a detailed study may be useful.

The effect of the shadow sector on the detection of targets in line with the obstruction depends, obviously, upon their size in relation to the width of the sector, and hence upon their range. In the average type of cargo vessel, say, with the scanner mounted above the bridge structure, the total shadow sector due to the foremast would usually be 1–3°, while those caused by the nearer samson posts may be 5–10°. In the case of the foremast there may be no *blind sector* but only a reduction of intensity and hence of the range of detection. It should be remembered, however, that there may not be sufficient intensity to obtain an echo from very small targets even at close range, despite the fact that a large vessel may be detected while considerably further away. The wider shadow sectors due

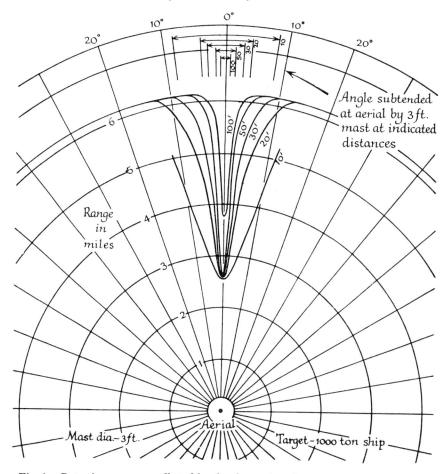

Fig. 62. *Detection ranges as affected by the obstruction of a mast at different distances from the aerial.*

to samson posts may have a blind 'core'. The funnel will usually cause a blind sector of from 10° to 45° or more, depending on its distance from the aerial. This sector will be large enough to obscure a ship of almost any size, however close.

Various experiments have been carried out to determine the shadow effect of obstructions to the radar beam. The results of these experiments have been combined in Figs. 60, 61 and 62 to show the effect of a variety of obstructions on the shadow areas caused. Fig. 60 shows the detection ranges of vessels of three different sizes as affected by the obstruction of a 3-ft. diameter mast placed 20 ft. from the aerial, subtending an angle of $8\frac{1}{2}°$. It will be seen that in the case of the 10,000-ton ship no shadow is caused until the range of 5 miles is reached; from that point the shadow sector gradually increases until it is about 24° wide at the normal maximum range of detection. It will also be seen that in the cases of the smaller vessels the shadow sector begins closer to the radar aerial. This is due to the lesser echoing areas of the smaller targets.

Fig. 61 shows the effect of varying the diameter of the obstruction while keeping the radar aerial at the same distance from it; Fig. 62 shows the effect of varying the distance between the obstruction and the radar aerial while keeping the size of the obstruction constant. In both these illustrations the target was a 1000-ton ship.

The differences between these practical results and the geometrical treatment are mainly that the angular width of the shadow sector does not remain constant and that it does not begin at the obstruction. These are both due to the fact that the radar beam does not start from a point but from an aperture which may be 6 ft. wide (i.e. width of the aerial).

The shadow sectors thrown by masts, samson posts and funnels are usually clearly discernible as dark sectors when sea clutter is visible on the PPI (Plate 31). When high gain is used these sectors tend to fill in. It should not be assumed that, because no clutter can be seen in these sectors, they are blind. All the foregoing considerations must be taken into account.

Measuring shadow sectors

Calculation of the probable size of the shadow sector from a knowledge of the width of the obstruction and its distance and bearing from the centre of the radar aerial gives a useful guide to its position on the PPI, but may be difficult if the shape of the object is irregular, as will be the case if there are cross-trees or steel ladders on a mast. There are two practical methods which may be used at sea: (1) observation of the blind sector against a background of light sea clutter, (2) observation of the bearings at which the echo of a small object, such as a buoy, disappears and reappears when the ship is turned.

When estimating the sector against sea clutter, the gain or sea-clutter control should be adjusted until the clutter is weak. Otherwise the sector may be narrowed due to clutter echoes penetrating into it from the sides. In the second method, when the ship is at a distance of a mile or so from a small object such as a second-class buoy not fitted with a corner reflector, she should be turned slowly through the appropriate arc, or through 360° if a complete survey is being made. The PPI should be carefully watched during the turn and the bearings between which the echo from the buoy is absent can be taken as indicating the extent of the shadow sector, or sectors. This test is best carried out in calm weather so as to avoid the possibility of the buoy echo being lost in the clutter or the buoy itself being hidden temporarily by waves.

False or indirect echoes

It has been shown that metal obstructions in the path of the radar beam may tend either to scatter it or to reflect it. Some structures such as funnels or cross-trees are exceedingly good reflectors and, having struck them, a considerable portion of the energy in the radio beam will be sent off at an angle which will depend on the character of the obstruction. It is necessary to recall that the radar beam has considerable vertical width and objects which do not obstruct the horizontal view from the scanner may give rise to reflections in the same way as those which do.

The echo caused by a reflected portion of the beam will return to the scanner by the same path and, whatever the actual bearing of the target may be, the echo will appear on the PPI on the bearing of the obstruction. It will appear at the true range of the target because the additional distance between the scanner and the reflecting object will be negligible. Such echoes are known as *false* or *indirect echoes*. The direct echo from the target will also appear on the PPI at its proper bearing and distance unless it is in a blind sector.

In practice, objects in the ship which lie in or near the horizontal path of the beam are the most frequent causes of reflections, and false echoes are, therefore, more likely to appear within shadow sectors. As has been seen, however, they may also appear on bearings where no shadow sector is apparent. Although false echoes may be caused by samson posts and less conspicuous structures the objects most commonly associated with these echoes are funnels and cross-trees. As only a small portion of the energy from the beam will contribute to the production of false echoes, it will require a target with a strong response to cause them. Thus the more usual false echoes are those of land targets, though they may also be produced by ships.

Assuming that the scanner is on the centre line, forward of the funnel, a false echo due to the latter will appear to be astern of the ship or nearly so. Owing to the curvature of the funnel the false echo will be so distorted as to bear little resemblance to the shape of the original target. When the ship is passing close to land or moving in a river, false echoes from the coast or the banks of the river may appear to be following the ship (Plate 32).

False echoes from ships are only likely to be observed when the screen is fairly clear of other echoes. The false echo of a ship approaching from ahead will, if reflected in the funnel, appear on the PPI at the same distance astern. As the range between the ships is being reduced at a rate equal to the sum of the two speeds, the 'ghost' ship will, if plotted, appear to be overtaking at that speed.

If the ship is sailing directly away from a coastline which has a good radar response, false echoes may be produced by reflections from the foremast

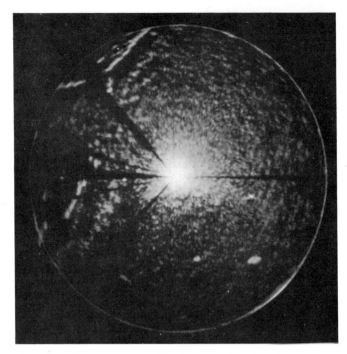

Plate 31. Shadow sectors. *The foremast, funnel and samson posts obscure sea clutter and land echoes.*

Plate 32. Indirect echoes. *Land near Sunderland is seen to the south-east due to reflection in the funnel; 3-mile range scale.*

Plate 33. False echoes. *The banks of the Scheldt, near Flushing, are reflected in the shadow sectors (same ship as in Plate* 31).

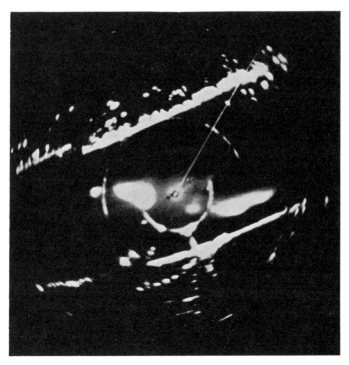

Plate 34. Side-lobe echoes. *The Humber near Hull; the echoes from the north bank and from the end of New Holland Pier are drawn out into arcs;* 1·8-*mile range scale. (The patches near the centre are due to reflection from the glass of the CRT.)*

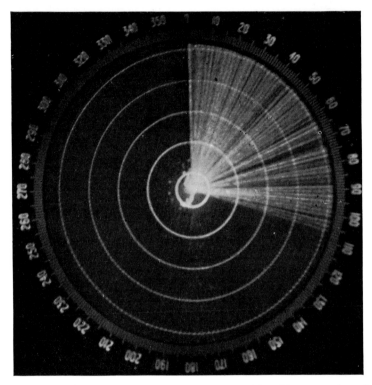

Plate 35. Spoking. *An example of serious spoking. Note serration of calibration rings.*

cross-trees (assuming that the scanner is abaft the foremast). These echoes will appear on the PPI to be ahead of the ship at the same range as the land and if plotted will appear to be moving at twice the speed of the ship.

A case in which appreciation of the possibility of false echoes is particularly desirable may arise when a ship is moving in a river in which strong echoes are received from the banks and from buildings on them. It will be remembered that when sea clutter is present shadow sectors caused by samson posts, masts, &c., show up as dark sectors in the clutter. At close range, when echoes are very strong, it is possible that all these obstructions may act as reflectors and cause false echoes to appear in the shadow sectors. If there is no sea clutter the shadow sectors may then appear as light sectors against a dark background (Plate 33). This may be slightly confusing, although of course the light portions will only occur at the ranges of any targets there may be.

These different examples have been quoted to show that false echoes may cause a certain amount of confusion and may even be mistaken for true echoes. It is possible to eliminate some of the sources of false echoes. In the case of cross-trees, for example, deflecting material of some kind (corrugated steel, for example) may be arranged either to scatter the beam or to deflect it upwards and so away from possible targets. Radar absorbent material (RAM) has also been found to be effective.

The characteristics by which false echoes may be recognized can be summarized as follows:

(1) They will usually occur in shadow sectors.

(2) Depending on the width and curvature of the reflecting surface, they will maintain substantially the same bearing relative to ship's head although the actual bearing of the target may change appreciably.

(3) Although on different bearings, they will appear at the same range as the corresponding true echoes.

(4) Their movements when plotted will often be abnormal compared with those expected of true echoes.

(5) Distortions in shape may be sufficiently obvious to indicate that they are not direct echoes.

It would not be entirely true to say that all false echoes are caused by reflections in the ship's own structure. When the ship is very close to buildings, bridges, dock walls, &c., these may cause false echoes in exactly the same way as has been described for the ship's structure. When the ship is in dock there may be such a confusion of false echoes due to buildings and the ship herself that it is virtually impossible to identify any particular echoes. A wall alongside a river berth has been known to act as a mirror to such good effect as to create a reflection image of targets on the other bank and in the river.

EFFECTS CAUSED BY THE EQUIPMENT

Side-lobe effect

It has been explained elsewhere that it has not yet been found possible to concentrate all the energy radiated by the aerial into a single narrow beam. Some of it is radiated as weak beams, known as *side-lobes* (see Chapter 4), which are sent out at various angles on either side of the main beam (Fig. 63). There is not enough energy in these side-lobes to produce echoes on the radar screen unless a target is close enough to give a particularly strong response. The echo from a particular target due to each side-lobe will appear on the PPI when that lobe is pointing at the target. The aerial at that moment will not be pointing at the target but the side echo must appear on the bearing of the aerial; it will be at its correct range. The usual effect of side-lobes, therefore, is to produce a series of echoes at the same range as the main echo but on either side of it.

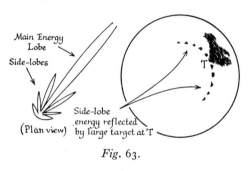

Fig. 63.

An example of this is shown in Plate 34. The echo from the end of the pier has been drawn out into an arc of a circle and a similar effect is visible with respect to the opposite bank. These arcs are practically semi-circles and while, in its severest form, side-lobe effect can cause a complete circle of echoes, it is usually confined to a small arc of echoes which rapidly diminish in size towards the extremities. Such side-lobe echoes can usually be eliminated by careful use of the gain and anti-clutter controls.

Set faults

Certain set faults manifest themselves not by a falling off in performance but by distortion of the radar picture or by other effects visible on the screen. It is therefore convenient to include in this chapter a description of the effects, though their causes will not be examined in detail.

Spoking. As is indicated by its name, spoking takes the form of a number of radial lines appearing on the display (Plate 35). These lines may occur all round the PPI or be confined to one arc, and may appear on each rotation or be intermittent. Spoking can be distinguished from interference from another radar by the fact that the lines are straight on all range scales, and are lines rather than a succession of regularly spaced pips. Should the trouble be confined to one narrow sector, it may resemble a ramark signal (Chapter 11), from which it may be distinguished by its steady bearing relative to the ship's head irrespective of the movement of the ship.

One cause of spoking is the presence of faulty or dirty contacts on rotating mechanism, such as the bearing transmission system.

Hour-glass effect. This describes a fault which is most obvious when a ship is in a river or very close to a coast. In a river, it is apparent as a constriction in width between the echoes of the banks near the centre of the screen. There is also the converse effect, in which the channel is widened out at the centre of the display, rather like an expanded centre. These effects can be caused by the emission of the pulse and the start of the time base not occurring at the same instant, or by a non-linear time base.

Sectoring of the display into alternately normal and dark areas is occasionally observed. No echoes appear in the latter areas. This usually occurs only on sets fitted with an AFC, and when this facility is very much out of adjustment.

Distortion of the picture when it first appears after the set has been switched on may occur with some cathode-ray tubes. Instead of filling the whole of the screen, the picture refuses to spread to the edge of the tube but behaves like a patch of water on a greasy surface. This effect, which is due to the build-up of static electricity inside the CRT, usually clears after a few minutes.

Serrated range rings are frequently displayed, and this fault is too often taken for granted by the user. Such serrations may be due to judder in the trace-rotating mechanism or to irregular action of the circuit forming the rings, and should not be present if the set is in proper adjustment.

The effects which have been described, though numerous, are readily recognized. Those indicative of set faults should not recur if the equipment is of good quality and is well maintained; those due to weather conditions, if they are not controllable by the experienced observer, merely demand caution, as do the permanent effects such as shadow sectors. No 'queer' phenomena should be allowed to pass into oblivion as mysteries; most of them will prove to be due to one or other of the causes mentioned, and recognition of them is the essential preliminary to their elimination or control.

8

RADAR AS AN AID TO NAVIGATION

RADAR has already shown how much it can contribute towards resolving the doubts and difficulties of navigation in low visibility and relieving the anxiety of those responsible for the safe and timely arrival of ships. However, if confidence in this aid is to be well founded it must be based upon a knowledge of its limitations as well as its capabilities. To use every available means of checking a doubtful position is the essence of careful navigation. Radar is but one of these means and, however obvious an interpretation of the PPI may seem to be, nothing should be allowed to obscure this fact. The radar set does not always produce clearly defined answers; its evidence requires interpretation, and interpretation needs skill, experience, commonsense and caution. Whatever claims are made in this book for the effectiveness of radar assistance are based on this understanding.

The significance of the information that radar can give as an aid to navigation naturally varies with the circumstances. These circumstances are best considered in terms of the particular requirements which give rise to them, which may be classified broadly as those of making a landfall after an ocean passage, those of coasting, and those of navigation in pilotage waters. The problems that arise in these three types of operation are therefore examined separately in this chapter, and the use of radar as a navigational aid in each is described.

The factors which vary most in the three problems are the accuracy needed in each and the time available for establishing the ship's position. Obviously the closer the ship is, or *may be*, to navigational dangers, the greater the accuracy which will be required and the less the time which will be available to make sure of it. The shorter the breathing space, the greater will be the benefit of experience. This, therefore, may be an appropriate moment to stress the paramount importance of practice in the art of radar-assisted navigation in untroubled clear weather periods, as well as the desirability of having the radar set always at short notice when approaching or in the neighbourhood of land.

LANDFALL

The caution with which a navigator approaches a landfall will depend upon the probable accuracy of his estimated position and the character of the coast he is nearing. It may be assumed that, especially if observations of heavenly bodies have not been possible for some time and no other recent fix has been obtained, the navigator will have taken full account of the probabilities of set and drift in making up his reckoning. Nevertheless, in many parts of the world there are possibilities of considerable unpredictable errors, which cannot be entirely eliminated by soundings. Although the latter may often give a reasonable position line or be employed as a clearing line for coastal and off-shore dangers they will in practice seldom give a fix from which the navigator may proceed with confidence. His main objective when approaching land in thick weather will be to identify, without any doubt whatsoever, some above-water feature and to determine his position relative to it.

Practical experience has shown that positive radar identification becomes easier the closer the ship approaches to general land areas, where more material becomes available for interpretation. At the longer ranges, when the actual coastline depicted on the chart is below the radar horizon, the identification of inland topographical detail with the echoes displayed on the PPI is a matter of some difficulty, particularly if previous radar experience in the locality is lacking. Charts are drawn mainly from the point of view of the appearance of the land to the eye, and landmarks which are conspicuous to the eye are not necessarily so to radar. As experience is gained and reported, however, more and more radar-conspicuous features are marked on charts. Previous chapters have described how the range at which a land feature may appear on the PPI will depend upon a number of factors, such as the height of the feature and of the aerial, the state of the atmosphere and the direction from which the land is viewed. Some preparation is therefore most desirable when intending to approach a landfall in thick weather with the aid of radar, so that the navigator may make some prediction of what should be seen on the PPI, and when.

The navigator should acquire a thorough knowledge of the capabilities and limitations of his radar. If he is wise he will study at the time those meteorological factors which will indicate the possibility of anomalous propagation, either favourable or unfavourable. Some ready means of indicating the range at which features of various heights will begin to appear above the radar horizon under standard atmospheric conditions is also essential. The following table (VIII), calculated for 3-cm. radar and standard atmosphere (see Chapter 5), shows the distance of the radar horizon from various heights. The values given will of course be affected by changes in temperature gradient, relative humidity and barometric pressure. They may be shown graphically as in Fig. 64 at (a).

The distance at which a feature will be on the horizon of a radar set is found by adding the distance of the radar horizon of the aerial to that of the feature. Thus, land having an elevation of 100 ft. will begin to appear above the horizon of an aerial mounted at a height of 50 ft., when the ship closes nearer than $9 + 12 = 21$ miles approximately. As has been pointed out in Chapter 4, it will need to rise above the radar horizon before it is likely to return an echo strong enough to paint on the PPI, the amount depending on the characteristics of the set and the echoing characteristics of the target (Fig. 49). The *least* height of an object which has appeared on the PPI at a range of 19·8 miles (aerial height 50 ft.), must be 80 ft. If the navigator constructs a curve such as that shown in Fig. 64 at (b) to suit the height of his own ship's aerial he will find it even more useful than the table.

TABLE VIII

Approximate Distance to Radar Horizon from Various Heights

Height (ft.)	Distance of radar horizon (n.m.)	Height (ft.)	Distance of radar horizon (n.m.)
18	5	215	18
24	6	240	19
32	7	265	20
42	8	320	22
54	9	380	24
66	10	445	26
80	11	520	28
95	12	595	30
111	13	680	32
130	14	770	34
150	15	860	36
170	16	960	38
190	17	1060	40

(3-cm. radar and standard atmosphere)

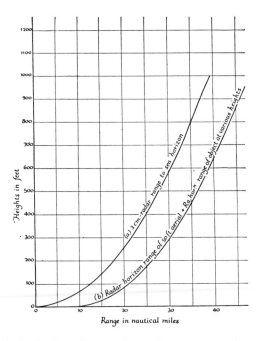

Fig. 64. *Radar horizon distances for various scanner and target heights.*

Radar position-finding at long range

Before seeking the first long-range echoes of the coast the navigator should of course assure himself that the character of the coast, within a reasonable distance on either side of his estimated position, is such that echoes may confidently be expected when the ship is still well to seaward of any off-shore dangers. Soundings provide in many cases a suitable means of increasing this assurance.

It is highly unlikely that positive land identification can be made if only a single long-range echo is painting on the PPI, unless it has a shape that is recognizable from past experience or it is a single high feature well separated from any others. A position obtained from less than three positively identified long-range echoes should not normally be accepted with confidence.

A convenient method of plotting a position from three or more prominent echoes is to transfer the echoes to a sheet of tracing paper and to plot the position by an adaption of the station pointer fix. From a point near the centre of the sheet, representing the ship's position, the echoes are plotted in true bearing and range, the latter being of course measured on the appropriate part of the scale of latitude of the chart in use. The tracing paper is placed over the chart, and correctly orientated to true north, with the centre near the D.R. position of the ship. The sheet is then moved about until all the plotted echoes correspond with charted features which may be expected to give responses. The ship's position can then be pricked through on to the chart. To facilitate this method of plotting, a transparent plotting sheet known as the radar station pointer has been produced as (British) Admiralty Chart 5028. Fig. 67 includes an example of a landfall fix.

It will be obvious that care is necessary before accepting identification of land features in this way. The navigator must be certain that the objects in question are above the radar horizon and are not shadowed by nearer features. In fact, he will be well advised to mark on the chart beforehand the objects which will be above the radar horizon at some given range during the approach. Whether or not this method can be used successfully will depend very much on the character of the coast. As has been said, previous knowledge, obtained perhaps in clear weather, of the objects in the locality which give long-range responses will also be a major factor.

In the approach to coastal regions which possess some useful long-range features there will be a period between the appearance of the early echoes and the moment when the coastline itself becomes recognizable, during which the picture may be confused with a large number of echoes which will be difficult to identify. If it is important to obtain a position as soon as possible, an effort to do so should therefore be made with the early echoes. Skilful use of the gain control may be of great assistance by accentuating the targets which are giving the strongest responses, but it must be remembered that these will not necessarily be from the highest features.

It is not always possible to obtain an accurate position at long range by radar ranges and bearings of a number of objects, and it is still less likely with simpler methods such as a radar range and bearing of a single object. It is not generally possible to estimate the nearest point of a slope or hill from which an echo will be received and, of course, the radar appearance and maximum detection range of a mountain or hill will often alter very rapidly for small changes in the bearing from which it is viewed. An *accurate* position is, however, frequently unnecessary at the landfall stage and even without a positive fix, information of the greatest value may be obtained by radar so long as all the prevailing circumstances are taken into account. It should be possible, for example, to make a landfall in the vicinity of Ushant and proceed up Channel with a very fair idea of the ship's position but without having made a positive identification of any particular part of the coast. The degree of accuracy needed will depend to a large extent on the character of the waters into which the ship is proceeding, and her subsequent progress will naturally depend to some extent on the margin of doubt as to the position, but the value of even approximate information is nevertheless considerable.

In parts of the world where other radio aids to navigation are available, their use in conjunction with radar may be very helpful. It may well be that in the vicinity of the landfall a good fix by Consol, Decca, Loran or direction finder may not be possible; on the other hand it is quite likely that any of these aids may give a position line which will limit the sea area in which the ship's position may be and so reduce the land area which has to be searched for echo-producing targets. A line of soundings may similarly be of value.

The question of the distance off the coast at which the prudent navigator will feel that he should know his position accurately is one which will depend on the circumstances of each case. The answer will usually be 'As far as possible.' A brief study of the table of horizon distances given earlier in this chapter will suggest that in many cases landfall targets will be well above the radar horizon at between 25 and 35 miles range.

Almost every volume of the Admiralty Sailing Directions (pilots) contains a table of the radar detection ranges of prominent land features. See Chapter 13.

A study of the meteorological conditions, employing only the instruments and weather reports normally available, may make it possible to take advantage of the presence of super-refraction and so to obtain a position further from the coast than is normally possible. When conditions are favourable to this the possibility of second-trace echoes should be borne in mind as these may otherwise tend to confuse. It is unlikely that second-trace echoes can be made use of owing to the difficulty in most cases of identifying the target. When the object is a lone island separated by several miles from other land, a second-trace echo might, however, be of some use if a rough position were needed. The least distance from a target at which a second-trace echo can be recorded depends upon the pulse

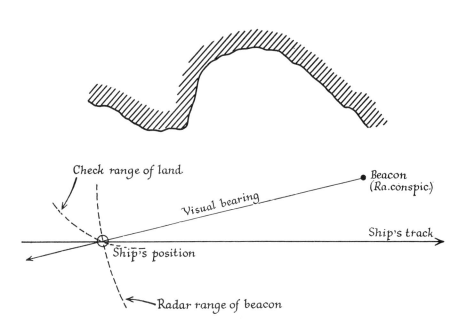

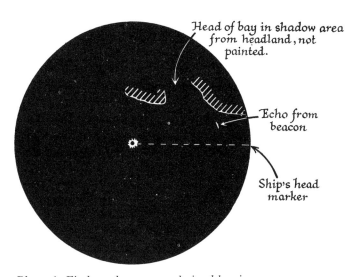

Plate 36. Fix by radar range and visual bearing.

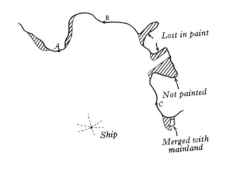
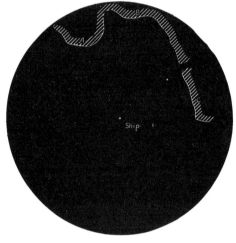

Plate 37. Fix by radar range circles, illustrating beam-width distortion.

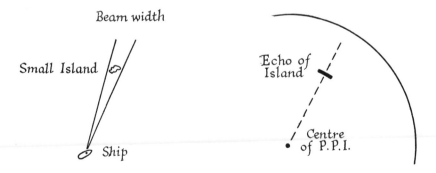

Plate 38. Estimating the bearing of an isolated target.

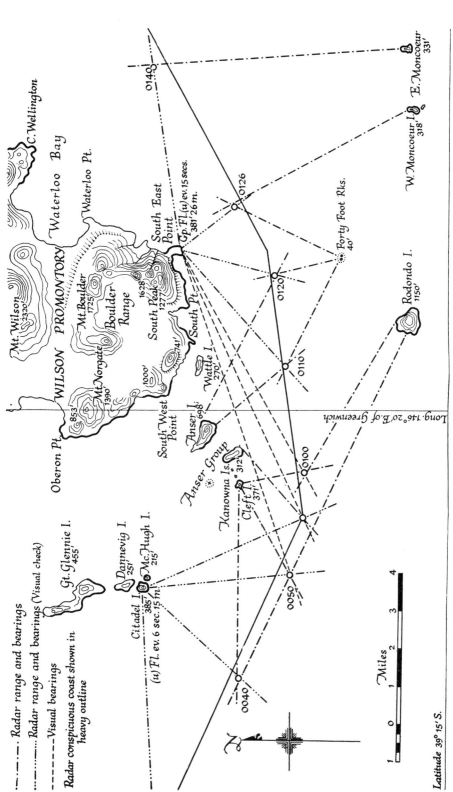

Plate 39. A record of radar fixing during a coastal passage (M. V. Wairangi).

repetition frequency. A table of least distances for different values of p.r.f. is given on page 125.

COASTAL NAVIGATION

Having achieved, perhaps, a reasonable position fix by long-range methods, the course will if necessary be adjusted and the ship will approach the land to within distances at which other radar methods may be employed. As has been said, the need for accuracy will be increasing and the navigator will therefore wish to plot the progress of the ship with precision. If the visibility is poor, radar may well be the only aid which will make this practicable. On the other hand it sometimes happens that a coastline lacks prominent features to an extent which precludes frequent visual fixing, even in clear weather. In these circumstances radar may again provide welcome assistance.

The range at which the radar appearance of a coastline will begin to resemble the delineation on the chart naturally depends on the topography in its immediate vicinity. It will seldom be less than 8 miles and frequently more than 15. A PPI picture which is clearly a paint of a certain portion of the coast may not always show particular features with sufficient clarity for a radar fix to be obtained. Even this response however may be a useful aid.

Before discussing the various ways in which radar information can be employed it may be well to emphasize the need to follow normal navigational practice in coastal waters. However simple and direct the radar evidence may be, the navigator cannot afford to neglect keeping a proper dead reckoning and estimate of his position and to check tidal set from any fixes obtained. Interpretation of the display is a matter in which the efficiency of individuals varies considerably, and the navigator will be well advised to read for himself any ranges and bearings he may require, even though another observer may necessarily be stationed at the PPI.

Methods of fixing by radar

In earlier chapters it has been explained that radar range accuracy is high whereas its bearing accuracy is low compared with that of visual observation. The descending order of accuracy of radar or radar-assisted position fixes may therefore be stated as:

(1) Radar range and visual bearing of prominent isolated objects.
(2) Radar ranges of several radar-conspicuous objects plotted as position circles.
(3) Radar range and radar bearing of a single charted feature.

It is here assumed (see below) that the range of a prominent isolated object can be more accurately measured than those of merely radar-conspicuous objects.

Radar range and visual bearing

Assuming that the characteristics of the radar set are satisfactory, the main consideration in relating a radar range to the chart is the determination of the point on the land which corresponds to the nearest edge of the echo. If a choice is available, an object which is high and steep-to in the direction of the observer should be selected; the radar range can then be related to the actual coastline on the chart with a minimum of error. If there is no choice, the navigator will have to take account of the amount of possible error, caused by the state of the tide and the composition and gradient of the shore. When this is small a very accurate position may be obtained by using the radar range in conjunction with a visual bearing, though a check by radar range of another object will always be reassuring. Such a fix is illustrated in Plate 36.

The method of combining a radar range with a visual bearing is of great value when coasting in weather which is thick enough to obscure all but the nearest parts of a coast or off-lying features. It can be used to good effect in clear weather when visual cross-bearings are impossible or difficult to obtain, e.g. when passing an isolated lighthouse.

Radar ranges plotted as position circles

When it is possible to select three prominent isolated objects, separated in bearing by, say, 45°, an extremely accurate fix can be obtained. Usually when coasting, however, the navigator's problem is to select suitable and identifiable portions of the paint of a coastline and plot their ranges as position circles. Since all parts of an indented coastline, except those lying at right angles to the observer's line of sight, will be to some extent distorted because of the beam-width of the aerial, care is necessary in selecting suitable points and in using the resulting observations. With 3-cm. radar the effect is not great but it should nevertheless be borne in mind when the choice of objects on which to range is made. The most suitable points on a coastline which lies at right angles to the line of sight will often be those nearest the ship.

The distorting tendency of beam-width is illustrated in Plate 37. The general effect is to broaden headlands, to fill in narrow indentations along the line of sight and to merge islands into the adjacent coast. The existence of shadow areas thrown over portions of bays hidden from the radar beam by high land adds to this effect. The general radar appearance of a strongly marked coastline will still resemble its actual form, but the detail will largely disappear and what remains will change considerably as the angle of view alters. As already remarked, portions of the coastline which are roughly at right angles to the line of sight will not be distorted and in the example illustrated, points A, B and C should be suitable from this point of view for fixing by range circles. The ranges observed are laid off on the chart with compasses from the three points, using the appropriate part of the latitude scale, and the circles will intersect or form a cocked hat at the ship's position. Naturally points should be chosen which are identifiable as accurately as possible and which are steep-to, so as to reduce to a minimum the errors due to the location of the source of the echo, and on this depends the advantage in accuracy of this method over the one which follows.

Radar range and radar bearing

The use of a radar range and radar bearing is listed as the method with the lowest accuracy of the three given because of the difficulties which have been referred to. It can be seen from the example shown in Plate 37 that it would be extremely difficult to take accurate radar bearings of the points A, B and C, from which the best range results may be obtained, even if the beam-width problem did not arise. This method will be of really practical value only when the target is either very prominent or isolated, or both.

When the target is isolated and of small extent it is possible to make a fair estimate of the bearing of its centre, and if it is steep-to, an accurate range can be obtained (Plate 38). The same applies to the bearing of a narrow promontory jutting out in the direction of the ship.

When taking a radar bearing of the side of a prominent land feature, a fair degree of accuracy can be obtained by applying a correction of half the beam-width to the observed bearing in the direction of the land. This is necessary because, as will be seen from Fig. 65, the edges of a headland cause an echo to paint while they are illuminated by *any portion* of the beam. When the edge of the beam is touching the headland, the edge of the echo on the PPI will appear on the bearing of the *centre* of the beam at that instant. The error will, therefore, be half the beam-width. As the effective beam-width is not a quantity which can be calculated with any accuracy it is not possible to give an exact correction, but the gain should be reduced when taking the bearing so as to reduce the error as much as possible. It should be noted that the direction of the extension of the land has nothing to do with the direction of rotation of the scanner.

It is necessary, therefore, to exercise considerable discretion when fixing by this method. In the example shown in Fig. 65 the ship's track will appear to run closer to the headland than in fact it does. This will only make for safety if there are no dangers close to the track on the other side. Whenever possible, confirmation of the position should be obtained by additional information such as a radar range of another point.

Radar range as a clearing line

In low visibility time is sometimes lost by giving a wide berth to prominent headlands when rounding them or when passing along coasts with off-lying dangers. If minimum range circles are drawn on the chart from the parts of the coastline concerned and clearing lines are drawn tangentially to them, the passage may be made in safety using radar ranges to ensure that the ship does not go within the minimum distances (Fig. 66).

It is, of course, essential if this method is to be used to make sure that the coastline will provide useful targets at the ranges needed and that contact is made with the actual coast or the targets selected and not with more prominent targets behind them. So that the nearest possible echo from the coastline will be seen, the gain should not be reduced below

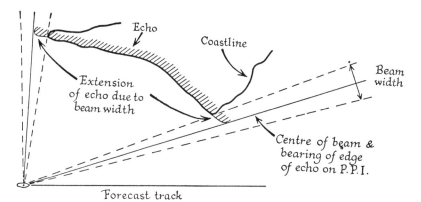

Fig. 65. *Correction for beam-width distortion.*

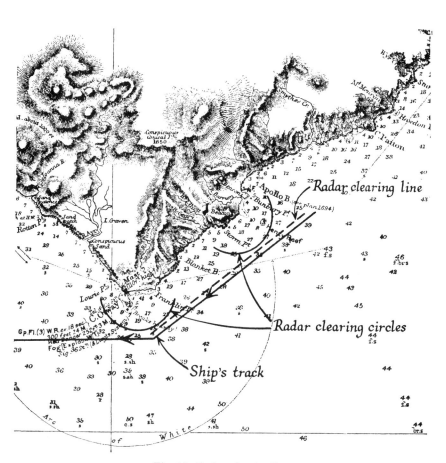

Fig. 66. *Radar clearing line.*

normal. This method of clearing dangers may be compared with the use of vertical and horizontal danger angles in normal navigation.

In good visibility the radar clearing line may be used to clear dangers lying some distance off a coast where visual fixing facilities are limited.

Coasting in general

The choice of which of the above methods to use will usually be forced on the navigator by the circumstances of the case, though any or all of them may be used in conjunction. Each of them carries its load of precautions which the prudent navigator will do well to study, but which with experience will become a second nature. Fig. 67 illustrates a landfall followed by a passage along a coast, in which each of the methods described plays a part; Plate 39 is a record of radar fixing carried out during an actual passage.

Very few situations can be imagined in which, with care and foresight, a coastal passage cannot be safely accomplished with radar assistance, whatever the visibility.

PILOTAGE

Having successfully made her landfall, approached the land to a safe distance and proceeded coastwise towards her port of destination, the ship will enter narrower waters, more thickly infested with dangers, though more plentifully provided with navigational marks, and incidentally more populated with others of her kind. In thick weather, the circumstances attending a landfall are varied, and even more so in coastal navigation; in pilotage waters, not only will the situations be of the widest variety but they will be changing from minute to minute. From the navigator's point of view the whole tempo of the voyage will be quickening, and at the same time the demands for accuracy in position fixing will be getting greater. The question of avoiding other vessels is dealt with in the next chapter; here it is only necessary to point out that necessary deviations from the intended track on this account will add to the exigencies of the moment and accentuate the need for speed and accuracy on the navigator's part.

At some point in this progress a great change will take place in the technique of navigation. In the earlier stages, time permitted the transfer of radar information from the PPI to the chart. In close pilotage waters, however, it may be assumed that the situation will change so rapidly, and alterations of course, for one reason or another, will follow so quickly in succession, that the plotting of the ship's track on the chart will have to be abandoned. The main function of the chart will then be to act as a guide to the identification of the land and to seamarks which the ship is passing or approaching. There will be occasions no doubt when a fix by one of the methods mentioned under Coastal Navigation will be possible and extremely reassuring, but they will be the exception rather than the rule.

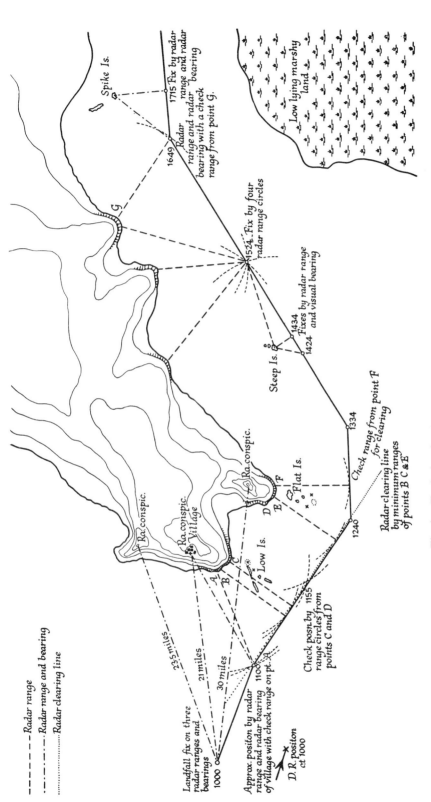

Fig. 67. Typical radar fixing during a coastal passage.

Presentation

The technique of blind or radar pilotage is dependent very much upon the method of presentation used. The question might have been studied under Coastal Navigation but it is of greater significance here. There are three methods of presentation of radar information:

(i) Relative motion, Ship's head upward;
(ii) Relative motion, North upward;
(iii) True motion, North upward. (See Chapters 2, 3 and 6.)

If the ship has no gyro or transmitting compass to stabilize the display, methods (ii) and (iii) will not, of course, be available.

Considering first the difference between Ship's head up and North up, the main issues are chart comparison and blurring. With the North up method, the PPI picture remains the same way up as the chart normally is and many contend that, amid the urgencies of a difficult pilotage, this allows the most rapid identification of marks and is greatly to be preferred. The ship turns, so to speak, within the picture which remains still (in orientation) except for the heading marker which indicates the progress of the turn. It is quite often necessary during pilotage that the intended track (course to make good) shall be on a particular compass bearing of a fixed mark, the echo of which can be seen on the PPI. This may differ quite considerably from the course steered, and progress made in the required direction can be checked directly if the presentation is north-upwards. Plates 40 and 41 represent a PPI during a large alteration of course when using heading- and north-upwards presentation respectively.

When heading-upward presentation is used the whole picture turns with the ship. Thus, particularly when a rapid alteration of course is made, the echoes of all objects on the PPI move quickly round and will be blurred or smudged due to the afterglow effect of the tube. This will temporarily destroy the definition of the outline of land echoes and may at critical moments be disturbing to the pilot. During a quick turn, the picture may be painting over the still visible afterglow of another part of the area. This also may confuse. On the other hand when in a buoyed channel, if any land or other navigational marks which may be on the PPI are of no significance, heading-upward presentation may be all that is required. This kind of situation is, however, comparatively rare; in the main it will be found that reference from the PPI to a chart during a blind pilotage is liable to be either frequent or urgent, or both, and in these circumstances the north-upwards presentation will be of more value.

Comparing the two methods which use north-upwards orientation, Relative and True motion, the latter has two important advantages in pilotage waters; echoes of ships under way may be distinguished at once by their trails, indicating their true courses, while the echoes of stationary objects may equally readily be identified by the absence of echo trails.

Naturally, to achieve this, the speed and course settings of the True motion radar must be kept in adjustment to suit the 'set and drift'. (See Chapter 3.)

General considerations in pilotage

Pilotage is not an easy subject to define. Navigation is both science and art and this branch of it is more of an art. No amount of text-book study will bring a painter or a pianist to the level of genius. What might be called the 'know-how' of pilotage is a matter of almost automatic reaction to the promptings of repetitive experience. In less complicated words, the expert pilot knows his own domain like the back of his hand. He has passed the multitude of significant marks in his familiar channels a thousand times; he can tell the state of the tide from a glance at a buoy or a discoloration in the water, and the feel of the wind will tell him how late or how early slack water will be at the berth. All these indications and many more will guide his successive actions as he takes his ships up from the sea and down again; but they cannot easily be got on to the pages of a book.

There can be no doubt that the essence of successful blind or radar pilotage will be familiarity not only with the clear-weather appearance of the land and seamarks but also with the art of radar interpretation in general and the radar appearance of the locality in particular. Since fog will deny the pilot assistance of many of the signs and portents which, in clear weather, would influence each decision as the need for it arose, a prime essential will be a preliminary study of the prevailing conditions.

Preliminary survey

Before entering restricted waters it is highly important to study the chart in order that a picture may be built up of the radar assistance which may be available. With 3-cm. radar a good response from sandbanks may sometimes be obtained, so that the state of the tide must always be borne in mind. It will be valuable also to know beforehand what landmarks are available for obtaining radar fixes should this be necessary, though the accuracy to be expected from such fixes will seldom be commensurate with that required in narrow channels. Needless to say, the capability of the radar set is an important factor, and knowledge of the normal detection ranges of the types of target likely to be met will be of assistance. In Chapter 4 it was explained how the radar response of buoys varies greatly with their shape, a can buoy, for example, giving a much stronger echo than a conical buoy. Large pillar buoys of the continental type usually give an excellent response. A table of detection ranges is given in Appendix I, but the following average figures are of the sort which the navigator should carry in his head:

High focal plane buoy	9 miles
Radar reflector buoy	7 miles
First-class buoy	3–4 miles
Second-class buoy	2 miles
Sandbank	1–3 miles

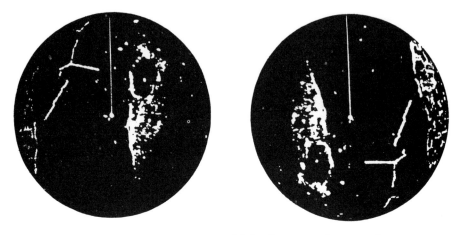

Plate 40. Two stages in a turn with heading-upward presentation.

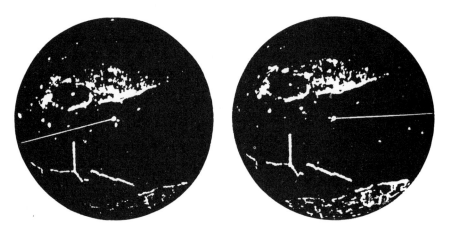

Plate 41. Two stages in a turn with north-upward presentation.

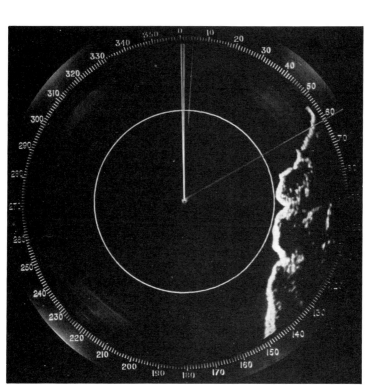

Plate 42. *Use of variable range marker to measure the distance off land; bearing cursor showing at* $61\frac{1}{2}°$.

These figures are for calm water and, since strong sea clutter may completely obscure the echoes of buoys not fitted with radar reflectors, particularly at close range, a study of the direction and force of the wind is highly desirable. Two factors of importance to the navigator are the minimum range of the radar set and the positions of blind sectors in its view. When passing close to a buoy it is reassuring to be able to see its echo all the time. The minimum range of 50 yards required by the British Ministry of Transport specification should permit this in most circumstances. When buoys in a straight channel are widely separated, interest in a buoy which has been passed may continue for some time. A ship which has only a small blind sector astern, or none at all, will be at an advantage in this respect.

Buoys and buoyed channels

Charts were drawn in the past to give the best service to the 'visual' navigator, though they are now being developed in directions which will increase their value when radar is in use. This is also true of the navigational marks on which the pilot depends. The increasing number of radar reflector buoys which can be seen, for example, round the coasts of Great Britain is evidence of this development. Reflectors are discussed in detail in Chapter 11, where it is shown that the object of increasing the detection range of otherwise poor targets can readily be attained. The problem of placing these aids so that they will be of maximum service to the pilot requires considerable thought. The advantage of increasing the detection range of buoys marking isolated rocks and shoals or those at the seaward end of buoyed channels some distance from land is obvious. The case of buoys marking a complicated network of channels, such as is found for instance in the Thames Estuary, is less clear. To solve this an understanding of the technique of pilotage is necessary.

Pilotage is necessarily a succession of moves from one mark to the next, and the pilot's interest in them will usually be concentrated on the mark he is passing and the one next ahead. If the marks are widely separated and considerations of cross-set or drift need to be taken into account, he may be interested in a mark for some time after he has passed it until the next one can be picked up. His immediate concern, however, will usually be limited to a radius of perhaps 3 or 4 miles. This applies in thick weather as much as in clear. The detection range on a good radar set of a first-class buoy is about 4 miles, and this should therefore meet the pilot's normal requirements in calm weather once his position is firmly established. In exposed waters it will no doubt be useful if some or all of them are fitted with reflectors to avoid their echoes being lost in sea clutter.

On the PPI, apart from possible differences in echo strength, one buoy looks very much the same as another. In negotiating a buoyed channel of any considerable length it is, therefore, most important to keep a careful check of the buoys which have been passed since, in very thick weather, visual identification may be impossible. The same applies to other marks which may be useful in determining the ship's position. In clear weather, if there is any doubt about the position, there will usually be sufficient identifiable land or seamarks in sight to fix it; this may not be so in fog. It is most important to identify positively and in good time the echoes of the next marks ahead and even the next but one, so that the ship is always moving into an identified area. Good warning is needed of sharp turns in the channel and of intersections or junctions with other channels, and at these points it may be provided if the buoys are laid in a recognizable pattern.

The technique of marking channels to assist low-visibility pilotage using radar may eventually be to employ ordinary buoys in straight or gently curving channels, with radar reflectors at intervals, and to provide more positive means of identifying such danger points as sharp bends or junctions. The proportion of radar reflector buoys needed will, however, be governed also by the severity of sea clutter to be expected.

In some places the normal arrangement of navigational buoys will form a distinctive pattern and these, or some of them, may be fitted with radar reflectors to increase their response. At a considerable distance, say 5 miles, from the pattern, with only the radar reflector buoys showing on the PPI, it may be quite distinctive. It should be remembered, however, that a check on identification may be required at quite short range, when all buoys in the vicinity will be giving strong echoes, and this may tend to destroy the distinctive nature of the pattern of the reflector buoys when this method is used. Careful use of gain control will assist in reducing the confusion of echoes. The reflectors can, of course, be mounted on fixed structures if this is more convenient. A useful method of identification is to note on the chart the range and bearing of a radar-conspicuous point from the buoy in question and then identify it by using the parallel index lines on the PPI cursor.

Leading lines (ranges)

When piloting by eye, the most sensitive aid to keeping a vessel in a narrow (straight) channel is the observation of leading marks (known in America as ranges). The slightest deviation from the leading line is evident when the marks are closely observed. Unfortunately, in the ordinary way radar does not lend itself to the minute identification of such small marks against a land background and, of course, radar bearing discrimination is itself an adverse factor. By using the parallel index lines on the cursor, it is possible to make use of displaced leading marks; if two such displaced marks are used, greater reliance is placed on radar ranges than on the less accurate radar bearings.

The possibility of using 'natural' targets as leading marks should not be lost sight of, though a great degree of familiarity with the locality is necessary before practical use can be made of them.

Other ships

The major problem in blind pilotage is the identification of echoes on the PPI. An aspect of identification with particular significance is that of detecting the presence of other vessels in the channel. Since this presents to the pilot a problem distinct from that of mere navigation, the longer warning he can be given, the better he will be prepared to meet it. It is no easy matter to pick out a ship target from a cluster of echoes, say 3 or 4 miles away, unless the radar appearance of the channel when unobstructed is very well known; if the visibility is very poor, the pilot will seldom have time to do it himself. If the other vessel is moving, it should be possible to distinguish her fairly rapidly by her change in position relative to echoes from identified fixed targets and, when using True motion, by her echo trail indicating her true course. If she is anchored it may be more difficult to do so and it should be remembered that in this case she is more liable to be troublesome. In either case it is certain that practice given in clear weather to the art of identifying targets of all kinds will repay handsomely in fog. A quick and easy way of clarifying a confused radar display is to plot the positions of doubtful targets on the chart by a parallel index range and bearing from a known fixed radar target, such as a pier or beacon.

Anchoring by radar

It is often necessary to anchor a ship in low visibility, and on occasions radar may be the only aid available. Although the position of the anchor on radar information may not be quite accurate, the assurance given by knowledge of the whereabouts of all shipping nearby will be of considerable comfort. The decision whether to seek shelter or to remain at sea depends largely on the suitability of the approach to the anchorage for using radar; no undue risk should be involved by the presence of other shipping as long as due account is taken of wind and tide and the way in which ships will be lying to their anchors. The approach course to the chosen anchorage should be made if possible with a radar-conspicuous object ahead; the range will then give a direct indication of the distance to go. Alternatively, the ship can be fixed by radar during the approach, using one of the methods previously described, or she can run in using radar clearing lines; parallel index lines on the PPI cursor will be of great assistance when using this method. In this case the approach courses should, where possible, be chosen so that they are parallel to the coast. An illustration of such an anchorage is given in Fig. 68. Used in this way

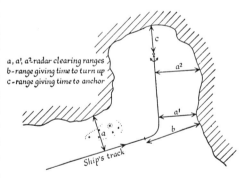

Fig. 68. Anchoring with the assistance of radar.

radar can also be of the greatest assistance to anchoring in clear weather, when visual marks are unhelpful or other ships are berthed nearby. Radar may also be used to great advantage in checking that, when anchored, the ship has not dragged. Again this method will be of greatest use in poor visibility.

Pilotage in general

It may be thought that too much complication is involved to make blind pilotage a generally practicable proposition. However, the most careful mariner may be caught by sudden fog in a narrow channel in which he is averse to anchoring. If his radar set is a familiar and tried friend, he should not have much difficulty in working his ship into less hazardous waters.

Radar undoubtedly offers great assistance to the master and pilot in fair weather and foul. It has been used in thick weather to enter and leave ports all over the world. The familiarity with localities which pilots attain should enable them to make rapid progress in this comparatively modern development of their art. Caution and attention to all the normal practices of navigation are, of course, prerequisites of success, as is also a preliminary study of the area and the conditions, should these not already be familiar.

Nothing has been said of the final stage of the voyage, berthing, because here radar, shore-based as well as shipborne, with its present form and characteristics falls short of requirements. There are many instances of ships being brought to their berths alongside in fog with the aid of radar, but these must be regarded as exceptional cases and have been made possible by great familiarity with the locality as well as experience in radar observation. A large measure of confidence and ability in radar interpretation is obviously necessary if the size of the ship is comparable with the area of safe water around her and the visibility is low. When all the visual aids to the detection and estimation of the movement of the ship over the ground are hidden, a two-dimensional plan on a scale of 4 inches to the mile is an imperfect substitute. It may be thought of as a 'bird's-eye view' from a height of 20,000 ft. Possibilities of overcoming this disability are discussed in Chapter 16.

Radar in small craft

With the rapid increase in the number of yachts and motor cruisers fitted with radar sets, which are invariably unstabilized and handled by operators who may be inexperienced in the practical use of radar, it is important that the situation does not revert to that of the 1950s when unstabilized displays were a contributory cause in a number of accidents.

It would be a great advance in the direction of safety if owners of small craft fitted with radar were to take a short radar course which would make them conversant with this aid to navigation, the like of which, correctly used, has not been seen since the discovery of the lodestone.

THE OPERATION OF THE RADAR SET

Other chapters have dealt with the composition and the operation of the radar set. Some of the points, however, are worthy of particular emphasis in connection with practical navigation and pilotage.

Range measurement. In some types of radar both fixed range rings and a variable range marker are provided, though the former are sometimes only intended for calibrating the marker; in others the marker is absent. Where fitted, the range marker should be used for accurate ranging because it obviates the need for interpolation between the range rings (see Plate 42, page 161). It is usually less accurate, instrumentally, than the rings, and it is therefore desirable occasionally to check the marker against the range rings. The accuracy to be expected after calibration is given in Chapter 3. To avoid either of these devices obscuring small echoes their width should be kept to a minimum.

Bearing measurement. An electronic bearing cursor is by far the most accurate means of measuring bearings, since it avoids both parallax and centring errors and the probably far greater yaw error to which unstabilized displays are liable. When measuring bearings with the mechanical cursor, care should be taken to avoid parallax error by viewing the PPI from a position directly in front of it. From the correct position the radial index lines inscribed on either side of the bearing cursor will appear as a single line which, when taking bearings, should bisect the echo. The centring should also be checked as described under Operational Controls in Chapter 3, and care should be taken to see that the centre has not moved significantly after an alteration of course. Errors due to the centre of the display being displaced will be at their maximum near the centre; therefore when accurate bearings are required a range scale which will place the echoes as far off the centre as possible should be chosen.

Heading marker. It will be unnecessary to point out to the experienced navigator that the heading marker does not necessarily indicate the direction of the ship's movement over the ground. To remember this is, of course, particularly important in river and estuary navigation, when strong cross-set may be experienced. It is used in setting the display to the correct orientation and so is of equal use at any time for checking that the orientation is correct. For this reason it is desirable to keep it switched on except for brief periods when making sure that it is not obscuring small echoes.

Performance monitor. Knowledge of the standard of overall performance which the radar set is giving is useful at any time and is particularly important when making a landfall and when entering restricted waters. A clear understanding of the function and use of the performance indicator or monitor is, therefore, desirable.

ERRORS AND EFFECTS

A few of the errors and effects which may affect navigation deserve mention in this chapter, though the notes which follow should not be taken to represent a comprehensive survey of possible radar errors.

Sources of error. There are many sources of error only some of which are under the control of the observer. All, however, should be kept in mind and given an appropriate place in any appreciation he may make. Some of the errors will be substantially constant while others may vary in random fashion or through the action of the observer. Generally speaking all of them affect navigational accuracy while only those which may vary between one observation and the next are likely to affect collision situations. They may be tabulated as follows:

Sources of error affecting navigation
  Heading marker displacement ⎫
  Compass error (permanent) ⎬ Bearing
  Beam-width error (p. 154) ⎭
  Uncalibrated variable range marker (pp. 54, 166) ⎫ Range
  Instrumental errors (see below) ⎭

Sources of error affecting any use of radar
  Compass error (random) ⎫
  Yaw error (only with unstabilized displays) ⎪
  Centring error (pp. 52, 166) ⎬ Bearing
  Parallax (p. 52) ⎪
  Using inner half of PPI ⎭
  Interpolation when using range-rings   Range
  Fundamental error of equipment, as stated by makers ⎫
  Not measuring to same point of echo each time ⎪ Range and
  Bad focus ⎬ bearing
  Gain too high (pp. 65, 154) ⎭

The instrumental errors referred to above concern range accuracy. They are unlikely to be large enough to affect anti-collision work. They are:

(1) *Non-linearity of PPI face.* This is due to the increasing curvature at the edges of the tube. It only affects this area and the error is usually extremely small. When measuring ranges, a range scale should if possible be chosen which does not place the echo near the edge.

(2) *Non-linearity of time base.* This is a defect which should not occur in high quality equipment. Its presence causes unequal spacing between the range rings and so should be obvious when its degree is of any importance.

(3) *Incorrect range-ring interval.* This again is an unusual defect. Its appearance differs from that in (2) because the intervals though incorrect

are identical. It is unlikely to be detected except by calibrating the set on a target of known range.

(4) *Index error*. This is due to the zero of the range rings not representing zero range. It may occur at the same time as (2). It will produce a different interval between the centre and the first range ring to that between the rings.

Errors due to (1) cannot be corrected. The removal of the causes of (2), (3) and (4) is best left to manufacturers or service depot staff.

Unwanted effects

A detailed description of unwanted effects is given in Chapter 7. In this chapter it will be well to emphasize three of them. (1) It is important that the observer should be aware of the existence and position of *shadow and blind sectors*. (2) Knowledge of the causes of and ability to identify *false echoes* may be of great importance when the ship is in narrow waters and shore and ship echoes are very strong. (3) Usually *side-lobes* occur only when in the vicinity of targets giving a very strong response. Their presence may greatly confuse the display and, if suspected, may be confirmed by reducing the gain. After identification the gain should be returned to the normal setting.

SHIPBOARD AIDS TO THE USE OF RADAR

Improvements to charts

A long bar scale of kilometres is shown in the outer border of Admiralty charts on scales larger than 1 : 100,000 to correlate with the calibration rings which may be shown on the PPI at these intervals. This is in accordance with a Technical Resolution of the International Hydrographic Organization, and a similar practice is followed on the charts produced by several other national hydrographic offices.

Although over the years there have been many suggestions for, and several experiments in, the emphasis on charts of features prominent on radar displays, the only permanent changes adopted have been:

(a) The use of a line of heavier gauge to ensure that the high-water line is more pronounced than on old charts, many of which are necessarily still extant;

(b) The inclusion of numerous height contours, and more spot heights, to the radar 'horizon' inland, replacing the representation of hills by the method of hachuring;

(c) The addition of 'Ra conspic' to identify certain features where this characteristic is known to be particularly noteworthy, yet where it might otherwise not be readily apparent from inspection of the chart; and

(d) The inclusion of radar responder beacons (Racons) where they are known to be being maintained, after evaluation, on a quasi-permanent basis.

The practice of showing by special symbol those buoys which incorporate radar reflectors has recently been discontinued, consequent on the introduction of the symbolization adopted for the IALA Buoyage System 'A'; this is because radar reflectors have now become usual in major buoys.

The policy of not providing special displays, for the radar user, of natural and cultural features on Admiralty charts is based on the following reasons:

(a) The insuperable problem of gathering and assessing such information on a world-wide basis;
(b) The practical difficulties of chart production and maintenance, which inhibit the use of a large number of colours and affect the amount of information that can be shown with the legibility required; and
(c) The very varying displays which result from differences in the type of receiver, height of aerial, distance from the feature and aspect of the feature.

In general, the clear depiction of features for general (visual) use equally serves radar use; and where some supplementary information on radar targets is called for, the present practice is the more economical one of incorporating it in the associated Sailing Directions.

An improvement in charting to assist ships in entering narrow waters or port approaches is the introduction of Radar Reference Lines. These are mid-channel lines on charts corresponding with lines incorporated in harbour radar displays. A line is used as a positional reference so that the harbour authorities may easily give a ship her position, relative to the line, when visibility is poor. In some cases the lines fall exactly on charted recommended tracks; the reference lines are then represented by superimposing the abbreviation Ra, in magenta, on the track symbols, at regular intervals. Where the reference lines do not fall on charted tracks, they are shown by a broken line, preferably in magenta, together with an appropriate legend and explanatory note on the chart. The special requirements of the local reporting and guidance system may require a reference line of particular design, e.g., it may be broken into sections of specified length and have reference names or numbers quoted. These lines are primarily reference lines and do not necessarily represent the exact tracks to be followed by all vessels guided by radar.

9

RADAR FOR COLLISION AVOIDANCE

(Note. *This chapter explains the basic principles of radar plotting and its simplest practical performance. A more detailed and advanced study is contained in the Annex.*)

RADAR has proved itself to be of very great value as an aid to avoiding collision in fog. The advantage which it confers, however, is not so much in assisting dramatic last-minute avoidance, as in enabling potentially dangerous circumstances to be rectified in their early stages. In studying the use of radar for this purpose it is most important to appreciate that radar is not at its best in rapidly changing close-quarter situations and that much depends on the manner in which the limited contribution it can make is used. In this chapter it will be shown that sufficient intelligence can be deduced from simple radar observations to construct the situation as it would appear to the eye in clear weather but that this cannot be done without a contribution from the user. To emphasize this point it is necessary to compare the efficacy of visual observation with that of radar.

Visual and radar observation compared

The circumstances of a daylight meeting with another ship in clear weather leap to the eye. The significant factors which are noted in the mind of the observer are the other ship's bearing and her aspect. *Aspect* may be defined as the relative bearing of own ship from the other ship. It is measured from 0° to 180° and expressed as 'left' (red) or 'right' (green) according to whether own ship is on the other ship's port or starboard side; if other ship is steering directly towards own ship her aspect is 0°. Continuous observation of the bearing will very soon establish whether there is a risk of collision and, combined with the aspect, it will usually form the basis for a plan of action, if any is needed, within a very few minutes of the sighting. When avoiding action is taken, any untoward movement on the part of the other ship which might prejudice the manœuvre's success will be immediately apparent if her aspect remains under close observation. The other ship's speed is not of particular interest in the general clear-weather case, since the avoiding action will presumably be taken while there is still plenty of sea room and its timing has no other significance. Nevertheless, to an experienced mariner, the relations between the other ship's aspect and bearing, and the movement of the bearing will give a good idea of her speed; this may be of value in special circumstances such as a close-range sighting or when manœuvring is restricted.

Although radar will give bearing information of the same kind and also the range, little else can be obtained directly from the PPI; only in exceptional close-range circumstances will the other ship's aspect be discernible; in fog, if ships are proceeding at 'moderate' speeds, echo trails will not be long enough to give accurate, quantitative information and there may be barely enough to suggest the directions of True or Relative tracks. The bearing may be seen to be steady or nearly so, indicating a risk of collision, but the degree of risk and its urgency will not be immediately evident. A slow movement of bearing and a rapid decrease of range may suggest the early development of a close-quarter situation while a similar movement of bearing with a gradual decrease in range may mean that the ships will pass well clear of one another. In clear weather the eye, noting the bearing and aspect, will have no difficulty in distinguishing between these extremes; radar information, however, cannot make the distinction with any certainty unless further details are deduced from it. When radar reaches a stage of development which will enable past information to be displayed automatically on the PPI in the form of True or Relative tracks or vectors, many of its shortcomings will disappear.

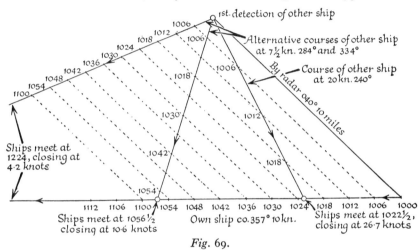

Fig. 69.

Fig. 69 shows how differences in the observed rate of closing imply widely differing aspects. In the example, own ship, which is steering 357° at 10 knots, detects another ship on a steady bearing of 040° with the range decreasing, and three possible situations depending on the rate of change of range are shown. If the other ship's speed is 20 knots, her course will be about 240°; if it is $7\frac{1}{2}$ knots, her course will be either 284° or 334°. In the first case the range will be closing at 26·7 knots and in the second case either 10·6 or 4·2 knots. Thus it will be seen that, depending on the rate of closing, the aspect of the other ship may vary by as much as 94° in the example taken. If the detection was made, for instance, at a range of 10 miles the ships would be very close after 22 minutes in the first case, while in the two others it would take either 56 minutes or 2 hours 24 minutes. One of the extremes in this example constitutes a leisurely overtaking situation, while the other is a highly dangerous crossing situation calling for lively appreciation and early action.

The rate of closing can be worked out from successive observations of range without any complicated arithmetic, but it is doubtful whether anyone but an expert in mental trigonometry would be able to deduce in his head the aspect of the other ship or construct a reliable picture of the circumstances. This is not to say that radar is useless without some form of computation; there will be many occasions when it will be impracticable and some on which it would be a distraction from more rewarding observation. Further, it is highly important to remember that plotting will only vouchsafe retrospective information; any prediction based on it may prove to be illusory. However, though it will not permit the observer to look into the future, it can arm him with a graphic and vital picture of the past and present.

The information required

The only safe and accurate way of getting the necessary information from radar observations is to compute them. It is quite feasible for this to be done automatically (see Annex), although it would be expensive, or to have some of the process automated. However, most observers will for a long time have recourse only to manual plotting. As an explanation of this will disclose the whole mystery of Relative motion, it will be the surest way to an understanding of the radar picture. Plotting is a simple and effective process which with practice becomes quick and easy to perform. Once its elementary principles are thoroughly understood, a familiarity with all the cases which will arise in practice can quickly be acquired through use.

It is not possible to define dogmatically all the information which a master will require and the sequence in which he will require it, since this will be largely a question of individual preference. However, in any potential collision situation certain items of information will be regarded as essential and these can be set out typically in different stages as follows:

(1) The time, bearing and range of the target on detection.
(2) The time (some minutes later), bearing and range; the direction of the bearing's movement and the direction and rate of movement of the range.

At this stage the master will be able to decide whether more information is needed. If it is, he will probably wish to know what degree of risk of collision exists. This can be ascertained from:

(3) The anticipated time, and the bearing and range of the other ship when she will be closest (CPA) if both ships maintain their present courses and speeds.

If this information gives cause for anxiety, the master will probably wish to visualize the exact circumstances. These can be given by:

(4) The aspect, or course and speed of other ship. Her latest range and bearing will of course also be passed.

Here the master will decide whether any avoiding action, either of course or speed, is necessary. He may then require information such as:

(5) A suitable alteration of course or speed, and the time at which to alter, assuming other ship maintains her course and speed.

If an alteration is made as a result of (5) or on the master's judgment, radar observation will be required to establish that the alteration is serving its purpose and that the other ship has not also altered in such a way as to defeat it. This may be given as soon as possible after the alteration by:

(6) A check on other ship's course and speed; a range and bearing and and indication of movement of range and bearing; a re-estimate on the lines of (3) and (4).

All this information can be obtained readily and accurately enough by plotting. Experience should enable the mariner to distinguish between variations due to errors of observation and plotting and those due to alterations of the vessels' movements. In connection with errors, the importance of using a compass stabilized display and an electronic bearing cursor has been mentioned in previous chapters. Regarding the use of a True motion display, it may be well to remember that although it may show with reasonable accuracy the course which the other vessel is steering, and although the bearing movement may show whether the other ship is likely to pass ahead or astern, neither True motion nor Relative displays will show directly how soon or how close the ships will come together on their present courses and speeds. This essential factor in estimating the risks of a situation, which is probably needed before any other, can only be obtained by plotting the movement of the echo or by computing it in some other way.

Before proceeding with plotting one further point will be made to emphasize the great importance of using a stabilized display. It is known that with some mariners, the predilection for the ship's head up picture is so strong that they will use it unstabilized even though stabilization is available. To take account of this bias, radars have been produced with a double stabilization. The picture is north stabilized on the C.R.T. and the C.R.T. is stabilized Heading up. These, though expensive, may be said to have the advantages of both methods.

PLOTTING METHODS

There are two main methods of plotting, either of which will give all the essential information required. The plot can either be constructed as a *True Plot*, which is similar in conception to a geographical plot, but need only take account of the movement of both ships through the water, or it can be constructed as a *Relative Plot* on which the movement of one ship

THE USE OF RADAR AT SEA

relative to another is plotted. The advantages of the relative over the true plot are that the information required can be extracted with less plotting and in the most appropriate sequence and that it is a quicker and more accurate method of forecasting the development of a situation.

To give reality to the examples, the results given in this chapter were obtained by actual plotting, to a scale likely to be used at sea. They differ slightly from those which would be obtained by calculation.

The true plot

In its simplest form the true plot is made as in Fig. 70. A line is drawn to represent the course being steered by own ship along which successive positions may be plotted on a suitable scale. Successive bearings and distances of the other ship obtained from the PPI are plotted to the same scale from the appropriate positions on the line. A line joining the other ship's positions will give her course, and her speed can be deduced from the time and distance between the positions.

In practice it is usual to mark off positions on own ship's course line at regular intervals ahead; since the distance steamed in six minutes will be one-tenth of the speed, this time interval is a convenient one. It is also simple to make the radar observations of other ship coincide in time with the intervals marked, so that they can be plotted directly.

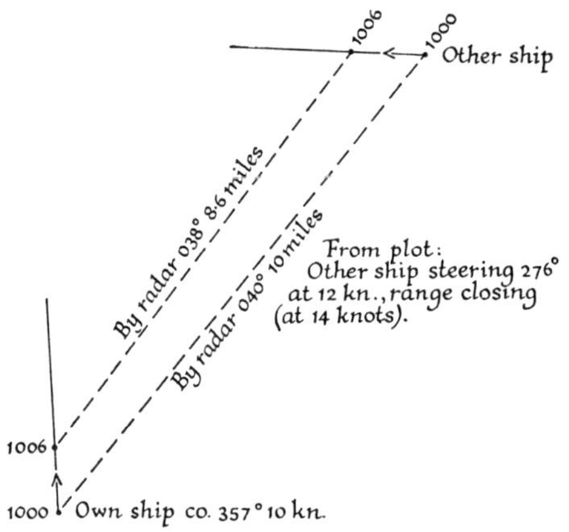

From plot:
Other ship steering 276°
at 12 kn., range closing
(at 14 knots).

Fig. 70.

In all the examples which follow, the plot is made and deductions are drawn on two observations separated by this time interval. In practice this would be unwise as the other ship might have altered course or speed between the observations. A third observation is essential to confirm the plot, and this would normally be made during the plotting interval.

It will be seen that the first information available from this plot is that required at stage (4) of the series mentioned above. A further construction is necessary to complete stage (3).

If the two course lines are extended, the anticipated positions of both ships can be marked on them, since the speeds are now both known. If there is any risk of collision the lines must intersect at a point which will be reached by the two ships at about the same time. A study of expected positions near the intersection will enable an approximation to be made of the time the ships will be closest together and of their bearing and distance apart at that time (Fig. 71).

This process discovers approximately the relative position which the other ship will be in when closest. It can be found more quickly and more accurately by first ascertaining the direction and speed of her movement relative to own ship, which movement is precisely that of her echo across the face of the PPI in relative motion. This construction can readily be done on the true plot, remembering that, to produce a relative picture, the successive bearings and distances must be laid off from a single position of own ship.

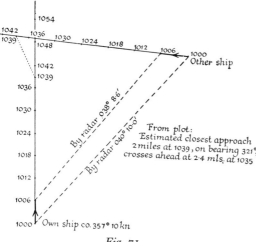

Fig. 71.

Continuing the plot from the point reached in Fig. 70, therefore, the 1000 observation would be plotted from the 1006 position (C) of own ship and marked O (Fig. 72); the 1006 observation already plotted is marked A (CO is equal and parallel to BW). The line joining them is then extended across own ship's course line cutting it at Z. The nearest point on this line to C is marked T, the angle ATC being 90°. The distance OA on the scale in use divided by the time interval between the observations (six minutes) gives the *relative speed* and the direction of OAT the *relative course* of the other ship. Her relative movement will be along this line at the relative speed; the time at which she will reach T, the point of closest approach, can be calculated and her distance and bearing at that time measured from the length and direction of CT. Thus the information required at stage (3) is obtained. That needed at stages (5) and (6) is best obtained by relative plotting methods.

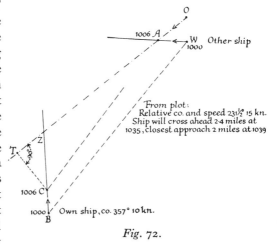

Fig. 72.

These are explained below under Relative Plot, but it will be seen that they can readily be applied to the true plot as a continuation of the construction in Fig. 72.

The relative plot

The relative plot deals entirely in terms of relative movement, the first principles of which have now been covered; it avoids having to plot the actual movements of either ship. The plot can, of course, be made, as was the true plot, on any suitable sheet of paper using a fixed point to represent the position of own ship; it is most conveniently constructed, however, on a plotting diagram such as that shown in Fig. 73. The reflection plotter and mechanical aids to plotting are discussed later in this Chapter.

On this diagram, which can be thought of as a PPI face, own ship is represented at the centre, the radial lines representing bearings and the circles, ranges. The top of the diagram (0°) can be taken to represent either north or the ship's heading. When the ship's head datum method is used, bearings are plotted relative to the ship's head. This has the serious disadvantage that when own ship alters course, all plots have to be transferred to suit the new relative bearings of the targets. For this reason in the examples which follow the compass datum method is illustrated.

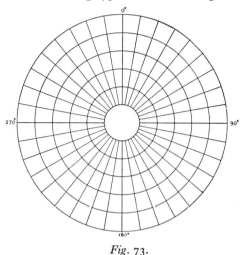

Fig. 73.

Speed-vector diagram. Referring to Fig. 72, it will be seen that a line drawn from W to O will be equal and parallel to BC.

It has now been shown that the length of each side of the triangle WOA represents a distance run in a given time, i.e. a speed:

WO represents course and speed of *O*wn ship;
WA represents course and speed of *A*nother ship;
OA represents course and speed of the other ship relative to own ship.

The first two of these quantities (WO and WA) are sometimes referred to as Proper Motion, and the last (OA) as Apparent Motion.

As the sides are proportional to the speeds they are known as speed vectors; the triangle WOA (Fig. 72) is known as a *speed-vector diagram*. If the other ship is stopped, OA will coincide with OW; W is sometimes called the zero-speed position.

As far as the plot is concerned, the six significant features of the triangle WOA are the directions and lengths of its three sides. If the direction and length of one side are fixed, only two of the remaining four features need to be known in order to complete the triangle. Thus if the courses and speeds of the two ships are known, the relative course and speed can be found; if the course and speed of own ship and the relative course and speed of the other ship are known, the latter's actual course and speed can be determined, and so on. In the method of plotting described here, the letters employed to define the speed-vector triangle have always the same significance, and the construction must permit the arrows to indicate truly the directions of movement. Thus the arrows on WO and WA must always point away from W; and the arrow on OA must always point to A, since it is towards this position that the other ship's movement, whether actual or relative has taken her. The relative movement is marked with a ringed arrow to distinguish it.

To construct the plot, two successive relative positions of the other ship are plotted at their observed ranges and bearings from the centre (C); these are lettered O and A respectively and the times inserted (Fig. 74).

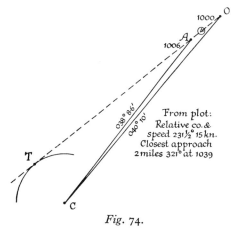

Fig. 74.

They correspond exactly with the points O and A used in Fig. 72. The *first* action on this plot, therefore, establishes the *relative* course and speed of the other ship. To find the point T, the line OA is extended and a circle is drawn (if one of the circles on the plotting diagram will not suffice) from C such that OA extended is tangential to it; the point of tangency defines T. The range and bearing of this point define the closest approach, the time of which may be calculated from the relative speed. Thus stage (3) is completed very rapidly and stage (4), to discover the course the other ship is steering, and hence her aspect, and her speed, may be proceeded with.

In the true plot the construction of the speed-vector diagram WOA was begun from the line WA, which indicates the course the other ship was steering and her speed. On the relative plot it has to be started from the line OA since the other ship's course and speed are not yet known. The principles of construction, lettering and arrowing are, however, exactly the same. The line WO, representing own course and speed, must be drawn from O in such a way that the arrow indicating the direction of own movement will point away from W. W may then be joined to A and the arrow inserted pointing away from W. The line WA, as in the true plot, represents the course being steered by the other ship and her speed. Hence her aspect may be determined and stage (4) completed (Fig. 75).

Planning the avoiding action. Stage (5) is to provide a plan for avoiding action. This may take various forms but it will be assumed as an example that the master wishes to prevent the ships closing to less than 4 miles

178   THE USE OF RADAR AT SEA

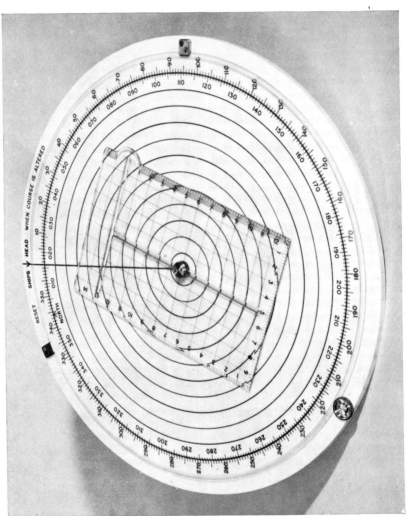

Plate 43. Compass-datum plotting device.

and that he will take avoiding action at 1012, i.e. six minutes after the last observation.

An expected position of the other ship at 1012 would be plotted at A_1 on the line OAT (Fig. 76), based on her calculated relative speed. An alteration by own ship at that time will naturally cause a change in the speed and direction of relative movement and the problem is merely to draw in the line along which relative movement is desired and then to discover what action by own ship is necessary to effect it.

A line is therefore drawn through A_1 touching the desired nearest approach circle at T_1 and extended backwards to provide one side of the new speed-vector diagram. As the course the other ship was steering and her speed were ascertained from WA in the original triangle and are assumed not to have altered, the line W_1A_1 may be drawn parallel and equal to WA. This establishes the point W_1. All that now remains is to draw the line W_1O_1 which will indicate the course and speed at which own ship must proceed so as to change the relative movement to the required direction.

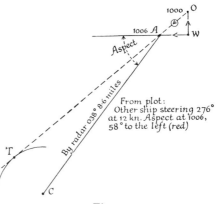

Fig. 75.

At this point a proper understanding of the constitution of the speed triangle is most important. In Fig. 76 the position of A_1W_1 has been fixed; point O_1 will, of course, be on the line A_1T_1 extended, but its position cannot be fixed until the direction of W_1O_1 is known. It will be seen from Fig. 77 that it is possible to draw W_1O_1 in a variety of directions, limited only by its length which, in turn, is limited by the speed available to own ship.

Any of the triangles $W_1O_1A_1$ would be a perfectly legitimate speed-vector diagram. When O_1 is between A_1 and T_1, however, the direction of relative movement is reversed because the arrow on O_1A_1 must always point towards A_1. Since this also involves own ship steering a course

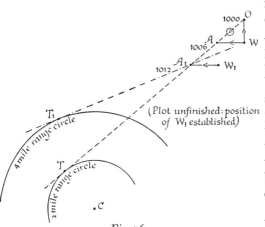

Fig. 76.

diverging from that of the other, there is clearly nothing to be gained by considering such a position of O_1 as a general possibility. The useful positions of O_1 will be somewhere on the backward extension of A_1T_1 and the choice will depend on whether it is desired to alter course or speed or both. As it is unlikely that own ship will wish to increase speed, it will be sufficient to examine the effects of an alteration of course and of a reduction of speed. The effect of altering both at once will then be evident.

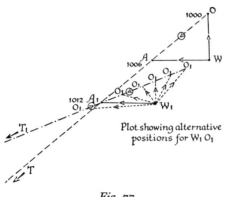

Plot showing alternative positions for $W_1 O_1$

Fig. 77.

To find the new course to steer at 1012 then, an arc of radius equal to WO (i.e. own speed) is described from W_1 (Fig. 78). Since in this example the other ship's speed is the higher, the arc will cut the backward extension of A_1T_1 in two places, either of which when joined to W_1 will indicate a course to steer to clear the other ship by 4 miles. Given sea room, it is usually better to take advantage of whichever course will cause the relative speed (O_1A_1) to be greater; in Fig. 78 this will be W_1O_1.

In the speed reduction case, the line W_1O_1 is drawn parallel to WO (own present course) (Fig. 79), and its length will give the speed required to effect the desired change of relative course. Any speed less than this will merely increase the distance of closest approach.

It will be obvious that any line drawn from W_1 to a point on the line between O'_1 and O_1 (between $W_1O'_1$ and W_1O_1) in Fig. 78 will indicate a practicable combination of course alteration and speed reduction which will achieve the desired result. Any one of the constructions described will complete stage (5).

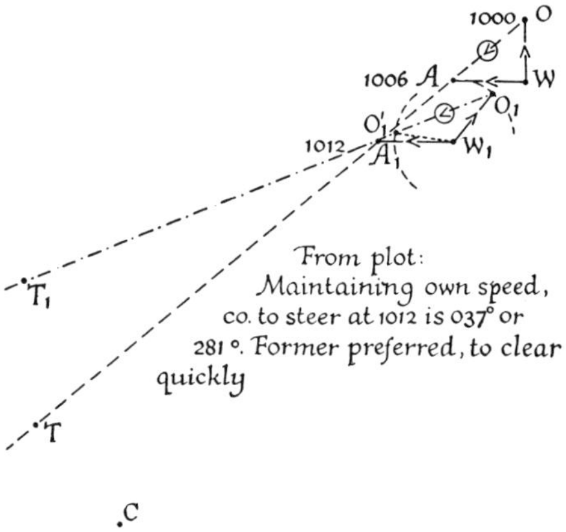

Checking the effectiveness of the action. The object of stage (6) is to discover at the earliest possible moment whether the action taken is going to be effective. During the period of an alteration of course by own ship, radar observations of the other may continue to be made and plotted, but it is unlikely that their significance will become clear, except to one highly experienced in plotting, until own ship is steady on her new course. The

From plot:
Maintaining own speed, co. to steer at 1012 is 037° or 281°. Former preferred, to clear quickly

Fig. 78.

state of affairs desired is that relative positions of the other ship after 1012 will plot along the line A_1T_1. Owing to the fact that own ship cannot instantly reach the new course and advance along it, these positions will appear on a line slightly nearer to own ship and slightly retarded in time. If a forecast of the new line is required a reasonable approximation may be made by assuming that the relative movement of other ship will continue beyond A_1 on the line OT for a period of half the time taken by own ship to turn to the new course. If, however, the alteration has been made in good time it will usually be safe to wait until own ship is nearly steady and then to take a radar observation followed by a second after, say, three minutes. If other ship has not altered course or speed, the line joining the plots of these observations (A_2, A_3) will be nearly parallel to A_1T_1 (Fig. 80).

It will, of course, be necessary to maintain close radar observation until the risk of collision has passed.

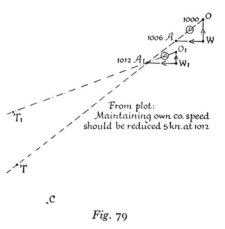

Fig. 79

If the master takes the avoiding action on his judgment of the circumstances disclosed by stage (4), stages (5) and (6) will not be applicable, but he will no doubt wish to ascertain that the action has been effective. In this case the plot will be continued from stage (4) by plotting fresh observations as soon as possible after the alteration and repeating the operation of obtaining the relative movement and the new closest approach and aspect.

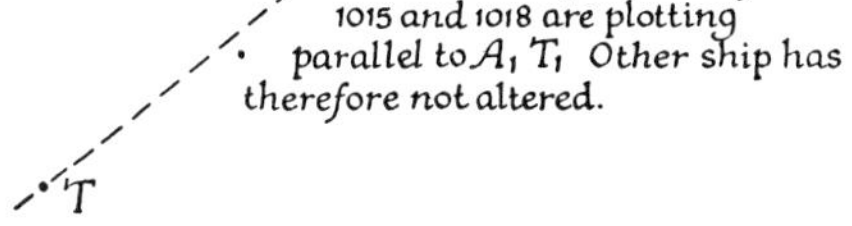

Effects of alterations of course and speed. Clearly, the direction of relative movement will be altered by changes either in own or other ship's movement. Fig. 81 will, however, serve to emphasize these facts. The changes at 1000 in the relative movements of ships W, X and Y are due entirely to an alteration of course by own ship at that time. It is particularly interesting to note that ship Z, which was on a steady bearing, reduced speed to 6 knots at the same time, with the result that her bearing remains steady and her relative speed has been greatly increased. Thus the effectiveness of own ship's action to avoid ship Z is negatived and the situation in fact worsened by Z's action.

Fig. 80.

182   THE USE OF RADAR AT SEA

Aids to plotting

Various devices are available for facilitating the plotting process already described and obviating the need to use drawing instruments, other than pencil and ruler. Five such devices will be described, the RAS plotter, the Decca ARP, the Kelvin-Hughes PRP, the Barr & Stroud (Parrish) Autoplot and the Reflection plotter.

The RAS plotter is a mechanical compass-datum plotter, designed by

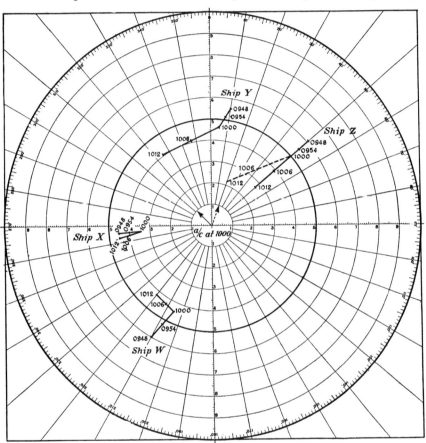

Fig. 81.

the erstwhile Radio Advisory Service of the Chamber of Shipping. It is illustrated in Plate 43. In consists essentially of a disc of transparent material, about 18 inches in diameter, which is engraved round its outer edge in degrees from 0 (North) to 359 and with concentric range circles from the centre outwards at intervals of about ¾ inch. This disc is free to rotate about its centre above a slightly larger circular sheet of material, which in turn is graduated at its outer edge in degrees relative, from 0 to 359, zero degrees being marked SHIP'S HEAD and the radius on that bearing being drawn in to represent the heading marker. The disc can be locked to the base and, the two bearing scales being contiguous, these can be placed in the appropriate relation to one another and locked there.

Attached to the axis of the plotter and free to slide over the face of the disc, is a transparent protractor, which is engraved with a grid, the lines of which are the same distance apart as the range circles. The protractor has a slot which permits it to move so that its edges can be used to draw lines on almost any part of the surface of the disc, and so that either of the two long sides can form a radius to the circle.

The plotter is set by placing the compass and relative scales in their correct relationship for the course being steered and locking them together. The distance between any two points is measured by moving the protractor until they are both on one of its edges and noting the distance between them on whatever scale has been chosen. The direction between two points is obtained by moving the protractor so that one long edge forms a radius, while the body of the protractor covers the two points, and rotating it until the long lines can be seen to be parallel to a line joining the points. The direction can then be read off from the end of the long edge.

Example. Own ship steering 330 at 10 kt.; echo bearing 040 at 8 miles at 1000, and $038\frac{1}{2}$ at 7 miles at 1006. With this range it will be convenient to use a scale of 1-mile between rings.

Lock the disc with 330 against ship's head. Rotate the protractor so that one long edge runs from the centre to 040 and mark point O at 8 miles from the centre. Rotate the protractor to $038\frac{1}{2}$ and mark point A at 7 miles range. Slide the protractor so that the points O and A are on the same edge; join OA and extend the line past the centre of the disc. Note the distance between this line and the centre (1 mile)—this is the closest approach; use the grid of the protractor to mark the point of closest approach (CPA); use a long edge to read the bearing of CPA; use the grid to measure the length of AT; as OA represents a time of 6 minutes, it is a simple matter to find the time for the echo to move along AT and hence the time at the CPA.

To find the true course and speed of the other ship, it is necessary to construct the speed vector triangle WOA. Place the protractor so that one edge passes through point O and is parallel to the heading line; draw a line from O in the opposite direction to own ship's motion and on it mark point W so that WO represents the distance run by own ship in 6 minutes. Join W to A; place arrows on WO and WA pointing away from W. Measuring WA in the manner described will show that the other ship is steering 278 at 12 kt.

Suppose that it is decided to alter course at 1012 in order to increase the CPA distance to 4 miles. Mark a point A_1 on the line OAT to represent the other ship's relative position at 1012. Move the protractor so that one edge passes through point A_1 and touches the 4-mile range ring; mark the latter point T_1 and draw a line T_1A_1 to extend back beyond A_1. Line A_1T_1 is the desired relative track. From point A_1 draw a line A_1W_1 parallel and equal to AW; with centre W_1 and radius WO describe an arc to cut the backward extension of T_1A_1; call this point O_1. W_1O_1 is the new course for own ship (058).

The Decca ARP (Automatic Relative Plot) is a semi-automatic means of drawing a relative plot. Like the RAS plotter, it gives a 'transferred plot', meaning that the bearings and distances have to be transferred from the radar to another device. However, in the ARP, the echo positions, selected individually by the operator, are transferred electronically by placing the bearing and range marker on the echo and pressing a foot pedal. They are transferred to a special plot console which has other special facilities for aiding plotting. One particularly useful aspect of this device is that the radar may be in True motion while the plot transferred from it is a Relative plot. Some of the advantages of both systems are therefore available.

The Kelvin-Hughes PRP (Photographic Radar Plot) is a special system in which the radar screen is photographed and projected on to a flat 2-ft. square plotting surface every few seconds. The basis of the plot is made by pencilling periodically the projected echoes directly on to the plotting surface. If the radar is in True motion, the plot will, of course, be a true plot. In this instrument the PPI itself cannot be seen, and the variable range marker and electronic bearing cursor cannot be incorporated.

The Barr & Stroud (Parrish) Autoplot, like the Decca ARP, is an electronically transferred plot of echoes selected individually and periodically by the operator. However, in this case the device displays true and relative plots of the echo simultaneously. The true plot is marked on a sheet of transparent material moving, according to own ship's speed, over a stationary sheet of red plastic which shows the relative plot. Own ship is permanently centred on the plot, which is stabilized course (intended) upwards irrespective of the mode of the radar, which may be stabilized or not.

The Raytheon TM/CA provides a reflection plotter and electronic markers on the PPI, which, with the echo itself, form the corners of the velocity triangle.

The reflection plotter is a device which is placed on top of the PPI, which is then viewed through the plotting surface and a 'semi-silvered' mirror. These are designed so that the echo can be plotted as if on the surface of the PPI and without any parallax. Obviously if the PPI is unstabilized and the plotting surface is fixed, a compass-datum plot will not be possible. To obtain the advantages of the latter, either the display must be stabilized or the plotting surface given a compass bearing scale and made rotatable so that the correct relationship can be maintained between the compass and relative bearing scales.

The reflection plotter has an advantage in that it obviates the need to measure ranges and bearings and transfer them to another plot; it has, therefore, a great advantage in speed and avoids some risk of error. Many experienced users consider that its provision should be mandatory. Great care in marking the echo is needed to avoid errors in the deductions.

When using a reflection plotter over a True motion display, sea-stationary, the result will be a true plot. To obtain the estimated closest approach it will be necessary to draw in the relative track. This can be done with probably the least disturbances and construction by moving own speed setting to zero, if this is possible. This will 'stop' own ship and from that moment the echo trails will be drawing relative tracks. These can be extended as soon as their direction can be discerned and the measurement made from the position of own ship. The same result can be achieved by construction on the plotting surface as shown in Fig. 72 or, more simply, perhaps, by erecting the side WO (Fig. 72), representing own movement in the time elapsed between points W and A, and drawing the relative track through points O and A. As own ship is moving on the plot, it would be necessary to mark its position at the time of observation for point A. The estimated closest approach will then be the distance from that point to the relative track. When using a reflection plotter over a relative motion display, the normal methods in Figs. 75 et seq. apply.

Other methods of computing

Hitherto the explanation has been concentrated on manual graphical methods of computing the radar information. There are various other ways of doing the job. Some of these, including fully automatic systems, are mentioned in Chapter 16 and dealt with in some detail in the Annex.

It has been shown above that two observations (preferably three) of range and bearing are sufficient to enable an estimate of the closest approach distance to be made. Table IX on page 190 gives this distance for any combination of range change and bearing change. Although it has been shown to be a simple matter to discover the course and speed of another ship from her echo movement by graphical methods, to do this by tables would be too complicated to be worth while.

THE PRACTICAL USE OF RADAR AS A COLLISION WARNING

When it becomes an habitual practice for users of radar in low visibility to note the bearing of any vessel sighted, and to continue to observe it until no risk of collision remains, the contribution of radar to the safety of navigation will have been greatly increased. The greater its regularity, the more effective will the practice prove in the relatively few cases where avoiding action is called for. Unless the habit is formed of plotting echoes and determining the aspects or the courses steered by other vessels, their passing distances and the times of meeting or crossing, mistakes may be made when they can be least afforded. Needless to say, regular examination of blind sectors is highly important. Recording of the simplest observations may prove vital.

Whether plotting is effective or not will depend upon the circumstances, including the time which can be spared for it. At a focal point, such as a pilot vessel's station, where ships in considerable numbers are reducing speed, stopping and proceeding again, full scale manual plotting will be useless; mere numbers of targets, however, need not necessarily make it ineffective, particularly if speeds are being maintained. The number of targets which may be kept plotted by one man depends very much on the individual, and estimates have differed by as much as four to one. It also depends on the facilities available. A reliable transferred plot from a display using ship's head upwards unstabilized presentation will need a team of two. On a reflection plotter, an ARP or a PRP and using a compass stabilized display, one man could easily cover many targets effectively. In neither case, however, will those concerned be able to carry out other tasks, such as keeping a visual lookout.

Targets to be plotted

If time is to be employed to the best advantage, the selection of targets for plotting is an important matter.

When in relatively narrow waters or following traffic routes, ships will be mainly on parallel or opposite courses, and it will usually be obvious which of them is worthy of close attention. In open waters and particularly at route junctions or crossings, it may not be so. If sea room is unrestricted and ships are converging from a variety of directions, a commendably bold alteration of course to avoid one ship may put a new complexion on the relative movements of others. A ship which has been harmlessly keeping station well out on the quarter may become a considerable danger if a turn of, say, 70° has to be made towards her.

It is a useful rule never to permit a ship echo to remain for long on the screen without making at least two observations of its range and bearing. Even if nothing further is done, the information thus made available may be of great value should later movements of own ship make it desirable to plot the other. Conversely, if there is an opportunity to plot the observations when they are made, and to establish the course the other ship is steering and her speed, time may be saved later when it is short. Naturally, attention will be concentrated on echoes which are closing on steady or nearly steady bearings, and the higher the rate of approach, the more urgently will they be considered. Thus, echoes of targets on ahead and bow bearings will be thought of more immediate importance but, as already remarked, it will be unwise to neglect all others.

The selection of targets for plotting will also depend on factors such as the visibility and the speed of own ship in relation to the speed of vessels likely to be met with. The slower the speed of own ship the greater will be the interest taken in echoes on after bearings. A reduction of visibility may render the use of radar desirable and yet be unlikely to cause ships to reduce speed much below their normal. In this case rates of approach may be high and the importance of early plotting will be increased.

A useful guide on the advisability of plotting radar targets may be:

Be aware of the bearings of *all* echoes on the screen and of their trends.

If the time permits, plot all of them, even if only to the extent of obtaining the relative track between two observations.

Always plot those whose compass bearings are nearly steady and whose ranges are closing, and obtain their courses, aspects, and the distances and times of closest approach.

Range scale to use

No seaman will regard the onset of a close-quarter situation in fog with equanimity. It has been shown that plotting takes time and that radar is not a rapid detector of change of movement. These factors and the knowledge that closing rates of up to 40 kt. are not unknown, even in fog, point towards the desirability of ascertaining the facts as early as possible and acting upon them in good time, and suggest the use of a long range scale.

If ship targets can be detected at 15 miles it would be wise to take advantage of the facility. At the same time it should be remembered that, although the same echo may appear on the PPI on more than one range scale, it may be more conspicuous on the shorter scales, while the better discrimination of the latter may disclose more than one target in a given area. On conventional radars these and other factors make it desirable to change the scale in use at regular intervals to ensure that all requirements are satisfied.

Assumptions about the other ship

The possession of radar by the 'other' ship is one of the unknowns which frequently enter into discussions on the use of radar. It is reasonable to assume in clear weather that the man on the other ship's bridge has the use of a good pair of eyes and a working knowledge of the Rule of the Road.

When ships are out of sight of one another, comparable assumptions would be that the other ship is or is not using radar, but either of these alone would be a dangerous basis for action. The first would involve further assumptions as to the efficiency of the set, competence of operation and intelligent use of the information. The second, the assumption that she has no radar, would invite unpreparedness for manœuvring action by her.

The only safe course is to allow for both eventualities and action, if taken, should be safe whether the other vessel maintains her course and speed or alters them. In any case unremitting observation will be necessary.

Occasions for use in open waters

Low visibility. At least one collision might have been averted if a radar-fitted ship had had this aid in use before she entered a fog bank. In one case on record a fog bank was sighted 6¾ miles ahead in daylight and the other vessel was struck ½ mile inside it. A proper caution will suggest having the radar set at 'stand-by' when there are indications of reducing visibility and commencing to operate in plenty of time to assist in retaining control of the situation. No doubt there will be many occasions on which the warning of fog will be much less than in the example given and it must be admitted that a vessel which enters thick fog at speed without previously sounding fog-signals is likely to be held to have contravened Rules 35 and 6 of the Collision Regulations. When any geographical or meteorological considerations suggest the likelihood of fog, readiness of radar for immediate use will no doubt be regarded as one of those precautions mentioned in Rule 2.

A collision between a vessel which was skirting a fog bank and one which came out of it is on record. In this case also there was held to be a contravention of Rule 35, but the value of the use of radar by either or both ships in such circumstances will be obvious. There are two distinct conclusions to be drawn. One is that it would be most unwise to cruise near to a fog bank without using any available means of knowing what may be coming out of it. The other is that, when using radar in fog, it cannot be assumed that all vessels whose echoes are on the PPI are also in fog and acting with corresponding caution.

Good visibility. Before leaving the consideration of collision risks in open waters, mention should be made of the possible value of using radar in clear weather. In areas of high traffic density and particularly those in which shipping routes from various directions converge, radar may be of great value in augmenting the information obtainable by eye. When danger is to be apprehended from more than one ship, a radar plot which will give the ranges and speeds of approach of other vessels may be of great assistance in resolving the relative urgencies of the situation.

Use in coastal waters

In coastal waters the PPI will be showing echoes of land or off-lying dangers, light-vessels, buoys and so on, in addition to those of ships. If, in these circumstances, time is available for plotting ship targets it may be useful to be able to forecast the arrival of other vessels into narrow channels or at turning points and thus minimize the risk of close-quarter situations. Other applications will no doubt suggest themselves.

At night in clear weather at congested points such as pilot stations, radar can be of great use in helping to avoid other ships and to discover the whereabouts of the pilot vessel.

Use in pilotage waters

Finally, a word must be said about collision risks in pilotage waters, buoyed channels, &c. The use of radar for navigation in such circumstances is described in Chapter 8. It should be mentioned here that over half the 'preventible' collision casualties have occurred in rivers and river entrances. In such waters the radar observer will seldom have enough time to make a comprehensive plot of ships. He will, however, be able to determine by using true motion or by very simple plotting which vessels are moving and which are not. When a vessel is ahead, or nearly so, and is anchored, her relative movement will be exactly the reverse of own ship's course and speed over the ground (i.e. including set and drift). If this is not the case, the other vessel cannot be anchored. When other ships are close to and about to pass or be overhauled, the radar observer probably can do no more than give a rapid and highly important succession of bearings and ranges.

In other chapters reference has been made to echo 'trails' which result from the afterglow or persistence of the cathode-ray tube. In close-range work, when plotting may be impracticable, these trails may be of great value in giving a rough indication of the *True or relative movement* or of a change in it. Afterglow trails should not be confused with the other ship's wake which may sometimes be seen against a sea-clutter background. The direction of the wake will, of course, indicate the course being *steered*.

Reporting from the plot

To avoid confusion, particularly when officers change ships, reports from radar observers, either from the display or the plot, should be made in a standardized form. Based on the six stages of reporting described earlier in the chapter and using the situation given in the examples, the reports might be made as follows:

Stage (1) 'Echo/Target Alfa zero four zero ten miles; probably a ship.'

Stage (2) 'Target Alfa zero three eight, drawing forward slowly, eight point six miles closing.'

Stage (3) 'Target Alfa CPA three two one, two miles in thirty minutes.'

Stage (4) 'Target Alfa zero three six, eight miles course two seven six' (or 'aspect five eight red').

Stage (5) 'Course to steer for CPA four miles from target Alfa will be zero three seven in three minutes time.'

Stage (6) 'Target Alfa zero two six, drawing aft rapidly, five point eight miles, course two seven six.'

Note. The target is identified by a letter from the radiotelephone phonetic alphabet. If bearings are reported relative to ship's head, they should be expressed in a three figure notation (000–180) and prefixed RED or GREEN.

Errors. A number of sources of error are under the immediate control of the observer and a number are not. In various chapters of this book reference is made to errors of range and bearing and to methods of keeping them to a minimum (p. 167). The observer can do little about the fundamental inaccuracy of the set, but he can do a great deal by careful centring, focusing, adjustment of gain, making sure that parallax error is reduced to a minimum and, if he is using an unstabilized PPI, by ensuring that the amount of yaw at the moment of taking the observation is accurately known. Failure to time the observations correctly may cause as much error as any of the other faults, such as incorrect interpolation, when measuring range by the range-rings, or inconsistent placing of the variable range marker or bearing cursor on a particular part of the echo.

It might be thought that the sum of all these possible errors would be insignificant, but this is not so. The effect of errors in bearing is most apparent when the rate of change of bearing is small. This slow rate of change may occur with a target in any direction, but when a ship echo is ahead or fine on the bow with a slowly changing bearing the effect of errors in the estimated course of the other ship may have a disastrous effect on the appreciation. When a ship is coming down from ahead or fine on the bow, the movement of the bearing in the early stages of the encounter will be well within the amount of the possible error. Hence, a converging course might be mistaken for a parallel course and vice versa, or a close-quarter situation for a safe passing distance.

At the range of detection, errors in range are unlikely to be as large as errors in bearing (compared linearly), but they may have a considerable effect when the echo is ahead or fine on the bow and own ship's speed is considerably the greater.

TABLE IX

Change of bearing during approach

Range factor	Closest approach factor														
	0·05	0·10	0·15	0·20	0·25	0·30	0·35	0·40	0·45	0·50	0·55	0·60	0·65	0·70	0·75
0·95		¼	½	¾	¾	1	1¼	1¼	1½	1¾	2	2½	2¾	3	3½
0·90	¼	½	1	1¼	1½	2	2¼	2¾	3¼	3¾	4½	5	5¾	6½	8
0·85	½	1	1½	2	2½	3¼	3¾	4¼	5	6	7	8	9¼	10¾	13½
0·80	¾	1½	2	3	3¾	4½	5¼	6½	7¼	9	9¾	13¾	14½	17	21
0.75	1	2	3	4	5	6¼	7½	9	10½	12	14	16	19½	25	41½
0·70	1¼	2½	3¾	5	6¼	7¾	9¼	11	13¼	15½	18½	22	28	45½	
0·65	1½	3	4¾	6½	8	10	12	14½	17	20	24	30	49½		
0·60	2	3¾	5¾	7¾	10	12	15½	18½	21½	26	33	53			
0·55	2½	4½	7	10	12½	15½	18	23½	28	36	56½				
0·50	3	6	9	12	15½	19¼	24	30	37½	60					
0·45	3½	7	11	15	19¼	24½	30½	39	63						
0·40	4¼	9	13½	18½	24½	31	40½	66½							
0·35	5¼	11	16¾	23½	31½	41½	69½								
0·30	6¾	13½	21½	30½	42½	72½									
0·25	9	17½	28	41	75½										
0·20	11½	24½	40	78½											
0·15	16½	30	81												
0·10	26	84													
0·05	87														

Range factor	0·80	0·85	0·90	0·95
0·95	4	5	7½	18
0·90	9½	12½	26	
0·85	17½	31½		
0·80	37			

10

RADAR AND THE RULE OF THE ROAD AT SEA

(1972 Rules)

WHEREAS in the 1965 Regulations for Preventing Collisions at Sea direct reference to radar was confined to the Annex on the use of radar information, the Regulations of 1972 have dropped the Annex and included its content and a good deal more in the text of the Rules. The object of this chapter is to draw attention to the principal instructions in which radar should play a significant part. These may be summarized as follows:

(i) Radar lookout.
(ii) Radar detection compared with visual sighting.
(iii) Proper use of radar.
(iv) Radar and safe speed.
(v) Radar and risk of collision.

In studying the Regulations it is important to note that all the basic operational instructions are grouped in Part B. Steering and Sailing Rules, in three sections, Section I (Rules 4 to 10) for any condition of visibility, Section II (Rules 11 to 18) for ships in sight of one another, and Section III (Rule 19) for restricted visibility.

Radar lookout

Section I (Rule 5) requires a proper lookout at all times by all available means appropriate in the prevailing circumstances and conditions. Apart from restricted visibility, there will be many occasions on which the use of radar will be appropriate; it has now become mandatory on such occasions. In addition to the somewhat general requirement for a lookout, Rule 5 indicates that its quality must be such that a full appraisal can be made of the situation and of the risk of collision. This is a new Rule of special importance.

Radar detection compared with visual sighting

The great differences between radar detection and visual sighting as a means of observation were discussed in the last chapter; the question really is whether 'in radar contact' can be taken to mean the same as 'in sight'. The Rules [Rule 3 (k)] define the phrase 'in sight of one another' as the state existing when one ship 'can be observed visually from the other'. So the Rule permits the assumption of mutual sighting in the visual case. It must be quite clear that no such assumption can be made in the case of radar detection. This, of course, is the reason why the Rules in Section II apply only to ships in sight of one another. In court cases ships have been faulted for using the sound signals for ships in sight of one another when they have radar contact only.

Proper use of radar

The proper use of radar covers a wide field of operational practice, only a small part of which can be dealt with here. The first point worthy of mention is perhaps the warning contained in Rule 7(c) that 'Assumptions shall not be made on the basis of scanty information, especially scanty radar information.' In the preceding clause (b) of that Rule the term 'proper use' is employed to cover scanning at long range to obtain early warning of risk of collision and carrying out 'radar plotting or equivalent systematic observation of detected objects'. Failure to use a long enough range scale has been a contributory cause of several collisions; an extension to this is distraction from the anti-collision function by using the radar, probably on a short range scale, for navigational purposes.

Radar and safe speed

It will have been noted that the expression 'Moderate speed' has been dropped in favour of 'Safe speed' and that a great deal more guidance is given on the considerations to be taken into account in choosing a safe speed to suit the circumstances than was previously offered. A number of considerations, of course, have nothing to do with radar; they concern the state of the visibility, of the wind, sea, depth of water and ship's draught, the presence of other ships or navigational hazards and the existence of artificial interference with the view from the bridge, such as bright shore lights or the back scatter of own lights from precipitation or other agency. The manoeuvrability of own ship is an important factor. [Rule 6 and 6(a).]

Rule 6(b) offers a list of additional factors which have to be taken into account by ships using radar. It will be worth giving these in detail:

(i) The characteristics, efficiency and limitations of the radar equipment.
(ii) Any constraints imposed by the radar range scale in use.
(iii) The effect on radar detection of the sea state, weather and other sources of interference.
(iv) The possibility that small vessels, ice and other floating objects may not be detected by radar at an adequate range.
(v) The number, location and movement of vessels detected by radar.
(vi) The more exact assessment of the visibility that may be possible when radar is used to determine the range of vessels or other objects in the vicinity.

The significance of most of these factors is self-evident. It may be worth observing that, in connection with clause (i) above, the controls available to the operator have a vital part to play, e.g. the range-scale as mentioned under Proper Use and also under clause (ii). Clause (v) contains what may be the most significant of the many factors which should be considered when deciding on a safe speed. Clearly the closing rate of another vessel may seriously affect the issue, while a sudden change in her movement may demand a revision instantly of the conclusion on safe speed. As defined, safe speed is such that the ship will be able to avoid collision and be stopped within a distance appropriate to the prevailing circumstances and conditions.

Radar and risk of collision

The 1972 Rules lay new and weighty emphasis on the need for a ship to be *constantly* in mind of the risk of collision whatever the state of the visibility (Rules 5, 6, 7, 19). In this connection all the factors listed in Rule 6(b) are of importance. Under clause (i), radar limitations, a significant item is knowledge of blind or shadow sectors and action to examine them periodically.

Rule 19(d) requires that a ship detecting another vessel by radar alone shall determine if a close-quarters situation is developing and/or a risk of collision exists. Rule 19(e) lays down 'Except where it has been determined that a risk of collision does not exist, every vessel which hears apparently forward of her beam the fog signal of another vessel, or which cannot avoid a close-quarters situation with another vessel forward of her beam. shall etc . . .' The great significance of these clauses is that they depend entirely upon appraisal of radar information for the decision on the manoeuvring action required. It has been remarked already under Proper Use of Radar that obtaining the needed information on risk of collision requires 'radar plotting or equivalent systematic observation'. It will be noted that neither the nature of the end-product of the plot nor the character of the alternative are explained. It will be realized that, if the radar appraisal results in a decision that own ship should move ahead or keep moving, there will be a heavy responsibility on the radar operator to monitor closely the movements of the other vessel and to report any adverse change immediately. The dropping of the phrase 'ascertainment of position' from the Rules [old Rule 16(b)] has removed a problem of interpretation, but the collision-risk problem remains for radar to solve; marine radar, alone or with manual plotting, is unsuited to dealing with swiftly developing close-quarter situations.

One further point in connection with close-quarter situations in which the fog signal of another vessel has been heard is the problem of identifying the radar echo with the ship whose fog signal was heard. If the radar is in first-class order and there is only one echo within fog signal range, identification one with the other may be thought established. If there are more than one echo in own ship's vicinity, she would be well advised to act with the greatest circumspection.

It could be said that the edge of a fog bank is an area of high collision risk, though it should not be if radar is being used. More than one collision case has been brought to court in which this fact was a contributory cause. Two which come immediately to mind are *Andrea Doria* and *Stockholm*, south of Nantucket, and *Grepa* and *Verena*, north of Cape Tenes, North Africa. In each case the latter ship was unaware of fog immediately ahead.

Action to avoid collision

This is governed by Rule 8 which says that it shall be positive and in ample time, if the circumstances admit. It requires the alteration of course and/or speed to be 'large enough to be readily apparent to another vessel observing visually or by radar. The action shall be such as to result in passing at a safe distance; the effectiveness of the action shall be carefully checked until the other vessel is finally past and clear'. In fog, these requirements obviously place a considerable responsibility on the radar operator and his equipment. (See above under Radar and risk of collision.)

Malpractices

Much food for thought may be found in a study of court judgments on collisions involving radar; these show many cases in which ships have been held to blame under one or other of the following:

> Failure to go at a Safe (Moderate) speed.
> Failure to navigate with caution when at close quarters.
> Failure to use radar.
> Failure to record observations.
> One ship in fog, the other not.
> Unwise radar settings: range scale too short.
> Lack of continuity of observations.
> Lack of intelligence in use of radar.
> Preoccupation with other duties, e.g. radar navigation.
> Misinterpretation of the radar picture and the need to plot.

Every one of these subjects has led ships into trouble; a chapter could be written on each, but it is hoped that to mention them here will encourage thought and action.

One of the most interesting cases was that of a collision between *Nassau* and *Brott* in which Rules 13 (Overtaking) and 5 (Lookout) were infringed. *Brott* was being overtaken by *Nassau* and had another ship fine on her starboard bow but crossing at a fine angle to port. The weather was rough and visibility poor. *Brott* altered correctly to starboard to avoid the danger from ahead and reduced speed, but she allowed her head to fall off too much in the trough of the sea and was run into by *Nassau* from the starboard quarter. For this *Brott* was found partly to blame, but *Nassau* was blamed for failing to keep clear and also for failing to observe the ship which forced *Brott* to alter to starboard, because her radar was on too short a range-scale. It is of particular interest to note that although the visibility was very poor in the area of the collision, the two ships had recently been in sight of one another, which caused the judge to invoke Rule 13, one of the Steering Rules.

Radar brings responsibility

The advantages which may be derived from the use of radar as an aid to avoiding collision have been discussed in some detail. In this sphere as in all others, however, privilege brings with it responsibility and it may be useful to conclude the chapter on this note. If the ship which possesses radar maintains it in a state of high efficiency, operates it competently and deals intelligently with the information made available there is no doubt that she can make a substantial contribution to safety at sea. An Admiralty Court has pointed out in effect that the use of radar must be intelligent if it is to excuse action which would be condemned in its absence.

From these two comments the generalization may be permitted that a ship whose equipment includes radar is given at the same time a considerable additional responsibility, which she will be unable to discharge unless:

(1) The set is kept in a high state of operational efficiency.
(2) It is used in all appropriate circumstances.
(3) The responsible officers are thoroughly conversant with its technical operation, its capabilities and limitations and the methods of dealing with the information obtained from it.

Conclusion

Finally, it may be useful to emphasize that action, including that based on radar observation, is required by the Rules:

(i) Rule 8(a): to be taken in ample time;
(ii) Rules 8(a) and (b): to be positive and large enough to be readily observed by another vessel visually or by radar;
(iii) Rule 8(d) to be such as to result in a safe passing distance and to have its effectiveness checked until the other vessel is finally past and clear.

These points may seem obvious and seamanlike but no collision comes to mind where they have been translated into action. An alteration, made early, should be bold enough to appear rapidly and unmistakably on the other ship's radar screen or plot. If it is less than 30° or 50 per cent. of speed it is unlikely to do so; many masters habitually alter 60–70° from a head-on approach. The term 'ample time' is given various interpretations; a useful way of approaching it might be to say 'emphatically not less than half the range at which the target was detected'.

It is not possible to overstress the importance of the third point. Unless its effect is closely observed, the most intelligent action to avoid may be turned into catastrophe by an unnoticed move by the other ship.

11

AIDS TO INCREASING ECHO STRENGTH AND TO IDENTIFICATION

MUCH has been said already about the poorness of some targets and the difficulties of identifying the echoes from many of them. In different chapters reference has been made in general terms to corner reflectors; they will now be discussed in greater detail and their construction and properties described. Two 'active' aids, the ramark and the racon, will also be described.

The considerable experience already obtained in the use of radar as a navigational aid has shown the need for additional information on the radar screen that will permit radar position fixing on navigational marks and on coastlines which of themselves are insufficiently characteristic in shape to give positive identification.

In some applications, assistance can be afforded by the use of passive aids in the form of corner and other radar reflectors, to increase the echo strength of targets with poor radar reflecting properties, such as buoys. In general, however, it is not easy to obtain unambiguous identification with this type of aid, while the cost of laying identifiable patterns of buoys is disproportionately high for the limited service they provide, and these are therefore little used today.

Broadly, the requirements for the active type of aid can be divided into three classes:

(1) A short-range aid to the navigation of estuaries and harbour approaches,
(2) A medium-range aid to coastal navigation,
(3) A long-range aid to landfall identification.

There is some divergence of opinion as to the extent of the information that is required from a beacon. One view is that to obtain a fix the beacon need only provide information which will serve to identify echoes already on the radar picture, whereas the contrary view is that the direct range and identity of the beacon should be obtainable from the transmission.

In some cases, where the navigation mark is situated on a coastline of reasonably characteristic general shape, one means of assisting in obtaining a fix is to provide on the PPI a radial line of bearing which will pass through the position of the mark. In this case, the identity of the mark can be determined from its position relative to an identifiable feature of the coastline and the range obtained from the intersection of the bearing line with the coastline. Where the coastline is featureless and the mark is situated either on it or away from it, additional information is required,

AIDS TO INCREASING ECHO STRENGTH   197

and in fact the usual active system employed today on any mark of navigational importance would normally provide that extra information.

Devices which could provide a radial line on a PPI, referred to as 'ramarks', were developed and in 1949 experimental equipments were established for a period of operational trials at Portland Bill and St. Catherine's Point on the English Channel. Later studies showed that a receiver/transmitter type of beacon, usually referred to as a 'racon', would be more suitable, and extended preliminary trials were carried out in 1954–5 on the Tongue and Bar Lightvessels. Production prototypes were subsequently installed at the Bar and Kish Lightvessels and at St. Abbs Head Lighthouse.

Since those days a considerable amount of development work has been carried out on these racons, with the result that a small, lightweight, low power-consumption equipment is now available which can be used on all types of navigation mark including buoys. A considerable number of these modern racons are now established world-wide.

RADAR REFLECTORS

The principles of the reflection of radar rays from various types of surface and their application to targets which have complex radiation diagrams were discussed in Chapter 4. The notion of *equivalent echoing area* and its dependence upon the inclination of the target has also been explained; targets of very small equivalent echoing area, such as buoys and small craft, will give weak fluctuating echoes, which will often be less than the strength of sea-clutter echoes and would in any case be detected only at short range even in the absence of sea clutter. This suggests a reason for the development of the radar reflector as a means of improving the radar response. In Chapter 4 the principles of reflectors were discussed but no account was taken of the fact that its own equivalent echoing area varies somewhat with the angle of view. These effects will now be examined.

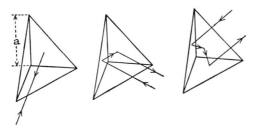

Fig. 82.

Two types of right-angle plate reflectors can be used, the dihedral and the trihedral. Dihedral reflectors, which are usually made up of two metallic plates at right-angles with the common side vertical, can be used to improve the echoing area of stable targets such as towers. However, the more usual form of reflector is the trihedral corner reflector used on buoys and small craft. In this, three mutually perpendicular plates intersect at right-angles, as shown in Fig. 82. These plates may be right-angled triangles or squares or right-angled sectors of circles. The fundamental property of the corner reflector is that, within certain limits of inclination, a ray entering the corner will be reflected back specularly in exactly the opposite direction (Fig. 82). This action can readily be demonstrated by fixing three good quality mirrors at right-angles to one another and either looking or shining a torch into them. Two everyday applications of the principle are found in the road signs which reflect the light from headlamps and so appear to be illuminated, and the 'cat's eyes' which are used to mark the centre lines of roads.

For any length of the short side (*a* in Fig. 82), which side is conventionally employed to define the size of a corner reflector, the square-sided corner has twice the physical surface area but nine times the equivalent echoing area of the triangular side corner. Unfortunately, however, the equivalent echoing area of the former falls off more rapidly than that of the latter for small changes of inclination, so that when wide coverage in azimuth is required, as is usually the case, the triangular corner reflector is to be preferred. For the same reason it is also preferred when a fitting on an unsteady object, such as a buoy, is required.

The maximum amount of energy will be returned to the radar receiver if the radar beam is directed into the corner reflector in such a way that the direction of propagation makes equal angles with all three surfaces; this angle can be shown to be 35°. At this angle the reflector will be displaying its maximum projected area to the beam. As the angle of incidence of the beam changes, the amount of energy reflected diminishes in such a way that, at about 20° from the line of equal incidence, the energy reflected back to the aerial is roughly halved.

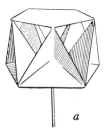
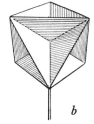

a   *b*

Fig. 83.

To ensure that the target will return good echoes from all directions the corner reflectors are assembled in groups, or *clusters*; the most usual arrangements being those known as the pentagonal and octahedral clusters. These are shown in Fig. 83 *a* and *b*; Plate 45 shows a pentagonal cluster fitted to a buoy. The pentagonal cluster consists of five separate corner reflectors mounted in a ring, each with one edge horizontal at the bottom.

AIDS TO INCREASING ECHO STRENGTH 199

The octahedral cluster is formed by three metal plates all intersecting at right-angles to make eight separate triangular corner reflectors.

Since the maximum amount of energy is reflected when the incident ray is on the line of equal incidence with all three sides of the corner reflector, thus permitting triple reflection, it will be desirable to tilt the reflectors to the appropriate angle. The maximum of the radiation from the aerial of a marine radar is in the horizontal plane, so, in the pentagonal cluster, the bottom surfaces of the reflectors are inclined at 35° to the horizontal. The construction of the octahedral cluster, however, is such that, when mounted

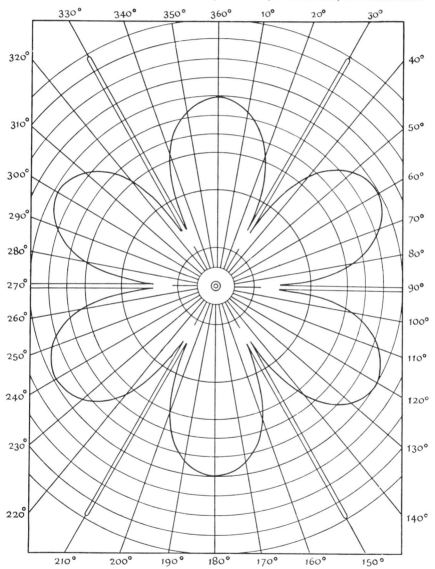

Fig. 84. *Theoretical polar diagram for an octahedral cluster.*

200   THE USE OF RADAR AT SEA

as in Fig. 83, the lines of equal incidence of the corners whose bases are uppermost will be 20° above the horizontal, while those of the others will be 20° below the horizontal. This arrangement is therefore a compromise which cannot take full advantage of the reflecting properties of the corners. In fact the maximum echo strength of an octahedral cluster is about half that of a pentagonal cluster using reflectors of the same size.

The polar diagrams for these assemblies are shown in Figs. 84 and 85, and are plotted through 360° in azimuth. They show one main lobe for each reflector in the cluster; evidently in the case of the octahedral, no

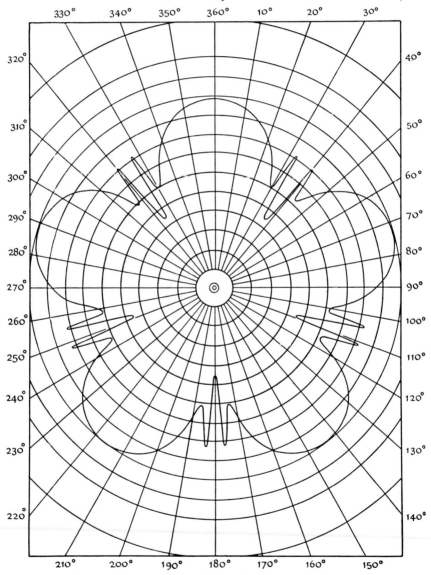

Fig. 85. *Theoretical polar diagram for a pentagonal cluster.*

response will be expected from the reflecting corners which point upwards or downwards except when mounted on a buoy which is inclined heavily by the movement of the sea. They also show that these assemblies do not give an ideal horizontal coverage, there being sectors where the reflections are considerably below the maximum. This problem may be mitigated to a certain extent by mounting a second cluster immediately above the first and displaced in azimuth by the appropriate angle.

New materials and new shapes

Although many corner reflectors are still made from metallic sheets, experiments have been conducted into the use of modern materials such as silver-coated nylon mesh embedded in glass-reinforced plastic sheets, and reflectors made of this material are now being introduced. They are much lighter in weight, can be moulded easily and are more resistant to corrosion. The nylon mesh also modifies the horizontal polar diagram to some extent, helping to effect some smoothing of the peaks and troughs in the response pattern, due to the multiple reflecting points in the mesh.

Other forms of radar reflector have also been produced. Some of these are based on the corner reflector, but with a modified shape; some use multiple clusters of small corner reflectors but others employ an entirely different principle.

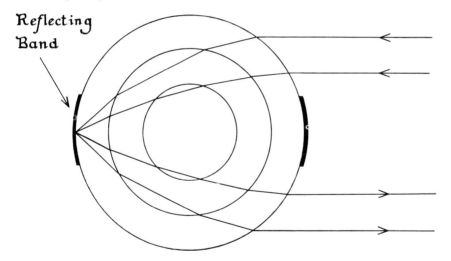

Fig. 86. Luneberg lens.

In the Luneberg lens (Fig. 86) a sphere is made up of a number of concentric layers of materials having different dielectric constants. A ray entering this lens is progressively bent at the interfaces of the different materials, as shown in the figure, until it reaches the far side of the sphere. If the dielectric constants of the materials are selected properly, the ray arrives at the far side of the sphere on its equator. By fitting a metallic strip around the equator, the rays are reflected as shown and return to the radar set along the same path as that by which they arrived at the reflector. The advantage of this type of reflector is that the horizontal polar diagram is theoretically constant over the coverage area. This area can be controlled by the amount of the circumference covered by the reflecting strip. If a full 360° coverage is required, the strip has to cover the whole of the equatorial circumference, but if only 180° is required, the strip will cover only half of the circumference.

The vertical coverage angle is related to the depth of the strip, but there is a limit to the dimension as the strip, being metallic, is impervious to the radar waves and the working area of the reflector is therefore reduced.

These reflectors are lightweight, easy to mount and less liable to damage by the waves than an octahedral or pentagonal cluster, but are more expensive and have a somewhat limited application. They are used in some countries though, in more sheltered waters, where they have proved very effective.

In some cases the manufacturer also applies a reflecting surface to the base of the lens so that it can be detected by overflying aircraft. This could have uses in search and rescue applications.

Factors affecting the use and selection of reflectors

As was explained in previous chapters, the effect of sea clutter on the display of a normal marine radar set is to produce a bright area at the centre of the screen which extends out to a distance determined by the roughness of the sea. This tends to obscure wanted echoes and, if a target in this area is a poor one, it may be obliterated.

Although the radar set includes a sea-clutter control, this will be useless unless the echo from the target is stronger than the echoes from the sea. The amplitude of echoes returned from buoys is often less than the mean clutter amplitude, even in moderate seas, and a radar reflector is required of a size sufficient to reverse this condition. From Fig. 87 it will be observed that a corner reflector of 17 inches side placed on a second-class buoy will overcome the effects of 3-ft. waves at all ranges but is effective against 8-ft. waves only at ranges greater than 2800 yards.

These curves also serve to illustrate the effects of the interference of the direct and sea-reflected rays described in Chapter 4. The minima encountered in this case at ranges of approximately 2400 and 5200 yards are representative of the reduction of echo strength which will be encountered when closing the range on a corner-reflector buoy in a moderate sea. When the sea is calm the loss will be greater and it is possible that the echo may disappear altogether when these points are outside the detection range of the buoy without the reflector.

From Chapter 4 it will be appreciated that, if the height and equivalent echoing area of a corner reflector are known, the detection range may be obtained from the coverage diagram (Fig. 41). Table X (page 206) gives

AIDS TO INCREASING ECHO STRENGTH 203

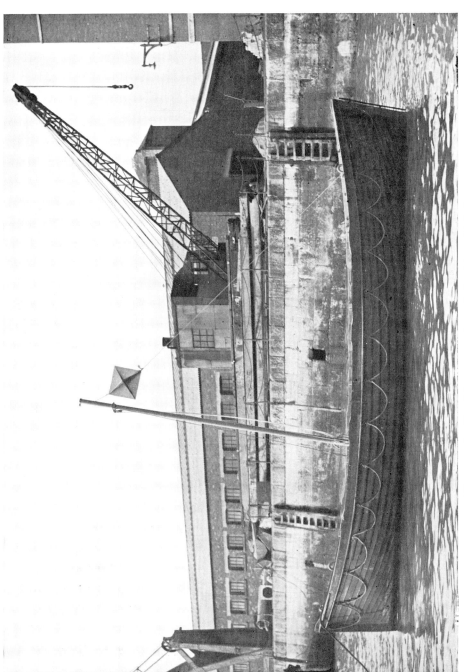

Plate 44. A collapsible corner reflector hoisted in a ship's boat.

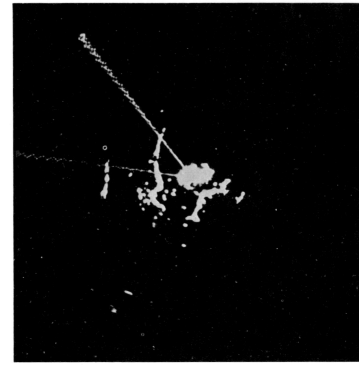

Plate 46. *Signals from two experimental ramarks; the ship is approaching Spithead and the ramarks are situated on land to the north and north-east.*

Plate 45. *A pentagonal cluster mounted on a buoy.*

Plate 47. In-band racon signal.

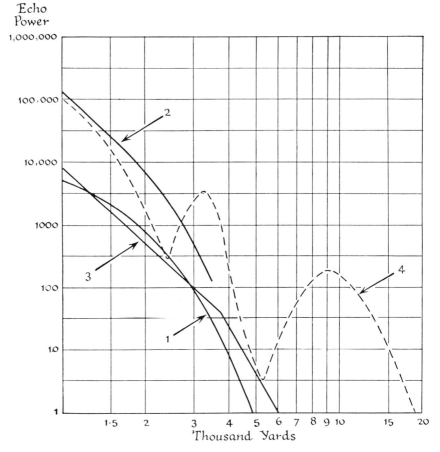

1 *Wave height 3 ft.*
2 *Wave height 8 ft.*
3 *2nd class buoy*
4 *17 ins. corner reflector on 2nd class buoy*

Fig. 87. *Comparison of echo strength of a corner reflector and of sea clutter at various ranges.*

TABLE X

Length of short side (ft.)	Equivalent echoing area (sq. ft.)
1	400
1·4	1,600
2	6,400
4	102,400
10	4,000,000

the equivalent echoing area of corner reflectors of various dimensions.

In comparison it may be noted that the equivalent echoing area of a second-class buoy is about 5 sq. ft. and of a 10,000 g.r.t. ship about 12,000 sq. ft. It will be understood that, if these figures are referred to the coverage diagram, they will apply to smooth sea conditions. In other circumstances there will be less reduction of strength in the minima, as was mentioned above, but the detection range will not differ substantially unless the object on which the reflector is mounted is affected by the movement of the sea.

Radar reflector identification

Many attempts have been made to solve the problem of the radar identification of buoys and marks by non-electronic means. Reflectors have been rotated and made to 'nod' periodically in an endeavour to produce an ON/OFF switching effect, but none of them has been really satisfactory. Practical difficulties such as those caused by the minima and the slow rate of the ship's aerial rotation militate against the success of such devices.

Practical aspects and applications of radar reflectors

Up to the present, the principal application of corner reflectors has been to increase the echo strength of such navigational marks as buoys and beacons so as to improve the range of detection and raise the echo above the level of sea clutter. The effectiveness of the reflector depends upon its height and size. Obviously there are practical considerations which limit the extent to which the equivalent echoing area of a target may be increased by this means. When the reflector is to be mounted on a buoy, for example, the desire to increase the size and height must be tempered by considerations of windage, buoyancy, the need for all-round cover and the mechanics of construction. Accuracy in the shape of the reflector is highly important; a displacement of 0·62 λ in the three plates will result in a reduction of the maximum echo strength to about one-tenth of that obtained with a true reflector. For a wavelength (λ) of 3 cm., this error is equivalent to about 2 cm. Accuracy of fabrication and rigidity, which inevitably implies somewhat heavy construction, are therefore essential. Consideration of weight, cost and simplicity has led to a general preference for the octahedral type of corner.

Statistical evidence has been collected to indicate the improvement that can be expected in the maximum detection range of a buoy when it is fitted with a radar reflector. These results have shown that the maximum detection range of a lighted buoy without radar reflector is in the region of 3·0 miles; by fitting a 17-inch octahedral cluster the maximum detection range is increased to about 5·5 miles. In the case of a third-class buoy the radar range is about 2·5 miles; the addition of a 17-inch octahedral cluster increases this to about 4·10 miles.

There are many other kinds of target, the normal detection ranges of which might be increased with advantage. Typical of these are wooden fishing vessels. Their normal range being perhaps an unreliable $2\frac{1}{2}$ miles, the addition of a corner reflector with a short side of 12 inches, hoisted on a yard or fixed to a mast at a height of 14 ft. might increase it to 5 miles or so. In such a case, however, it would be well to remember that the echo strength is much dependent on vertical inclination, so that a steady echo would still not be expected from such a craft in a seaway.

Similarly, the addition of this device to small boats, particularly lifeboats, might be of great assistance to ships attempting to avoid or to locate them. Aids of this nature were supplied to lifesaving rafts and dinghies during the 1939–45 war. Since then a great deal of thought has been given to the problem in various countries. Various forms of collapsible reflectors have been produced. The dangers which yachts undergo when sailing in open waters have led in some cases to insistence on the provision of reflectors. The Inter-Governmental Maritime Consultative Organization (IMCO) has issued a recommendation for the operational requirements covering radar reflectors on small craft. A simple and robust design for a

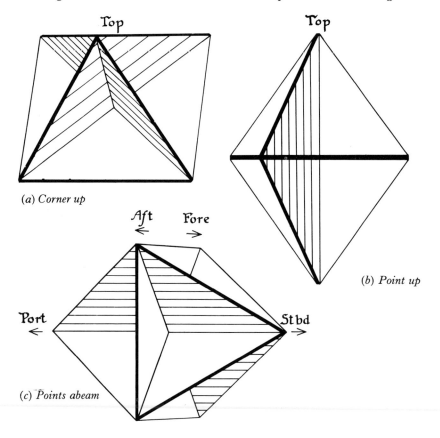

Fig. 88. *Octahedral reflector.*

collapsible corner reflector for use in such circumstances is, however, shown in Plate 44. This can be rapidly erected and hoisted to the boat's masthead.

The mounting of a radar reflector on a boat is also of importance. It has already been mentioned that with the octahedral reflector, there are six operating corners in azimuth. This is only obtained if the cluster is mounted correctly as in Fig. 88(a). Unfortunately many reflectors on boats are mounted from a corner of the cluster, as shown in Fig. 88(b), as this is easier to accomplish. This mounting can, however, seriously affect the effectiveness of the radar reflector and should not be employed. As a compromise, though, the mounting shown in Fig. 88(c) can be used, in which one plate only is vertical with its top edge horizontal and on the fore and aft line. This is particularly useful on a yacht, since, as it heels, the reflector tends to take up the position shown in (a).

RADAR BEACONS

It has been shown that although the passive radar reflector may be used to improve the response of a navigational mark very considerably, it will seldom in itself afford identification. When this is needed, the mark must exhibit some individual characteristic on the radar screen. The navigator will be interested in fixing his ship's position; hence his total requirement in this case will be ranges and/or bearings, and certainty of the positions from which these observations are measured. As stated in Chapter 8, a position may be determined by range circles from identified points; it may also be fixed by two accurate bearings or by a combination of range and bearing. It has already been implied that the amount of information required from the mark itself will depend upon the amount otherwise available from the radar screen, that is to say upon whether the adjacent coast has characteristic features. Whatever the balance of requirements may be, the mark must exhibit on the PPI some characteristic other than a simple echo and, therefore, some form of radio signal which will preclude ambiguity must be transmitted by it.

Radar beacons may be classified into two types, the in-band, which operates within the radar band, and the cross-band, which transmits on a frequency outside it. In-band beacons may also be of two types, swept-frequency and fixed-frequency.

The operating frequency of a radar set is determined by the frequency of its magnetron, which could be anywhere within the appropriate radar band (2920–3100 MHz in the 10-cm. band, and 9320–9500 MHz in the 3-cm. band). If a beacon is to be received without using additional equipment in the radar set, its transmissions must cover all of that part of the band covered by the radars. There are various ways of achieving this wide frequency coverage and the broad-band beacons in use at the present operate on the principle of continuously varying the frequency between certain limits.

In-band fixed-frequency beacons, or band-edge beacons, transmit on a fixed frequency at the edge of the radar band. A band of 20 MHz (2900–2920 MHz and 9300–9320 MHz on the 10-cm. and 3-cm. bands) is currently set aside for these beacons. In order to receive their signals, the local oscillator of the radar set can be retuned and use made of the existing IF and video amplifiers, or a separate receiver, fed from the radar antenna, can be employed and the signals applied directly to the radar display.

The cross-band beacon transmits on a fixed-frequency outside the radar band. Here again, the radar local oscillator could be retuned if the frequency was close to the radar band, but if the frequency was well separated from it, an independent receiver, which could be connected to the radar display, would be required.

If use is made of the radar receiver to process the signals from either of these fixed-frequency beacons, the receiver will not accept radar signals when tuned to the beacon frequency, so only the beacon signals will be displayed on the screen.

IN-BAND RADAR BEACONS (RACONS)

The in-band radar beacon (racon) can be considered as a form of electronic reflector which will return to the radar set a signal which is not only magnified but which has unique characteristics so that its appearance on the radar screen will readily identify the object from which it comes. The principle of the racon is shown in Fig. 89. The pulse of energy which leaves the

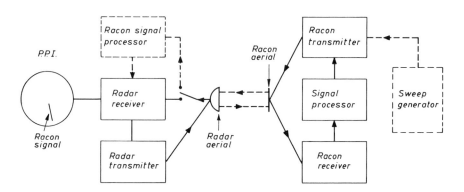

Fig. 89. The racon principle.

radar aerial is received by the racon aerial and fed to the receiver where it is amplified. The resultant signal is processed and used to operate the transmitter which radiates the characteristic signal. Sweep circuits are used to change the frequency of the transmitter in the swept-frequency beacons. At the radar set, the signals are detected either by applying them directly to the radar receiver, in the case of the swept-frequency beacon, or by processing them by retuning the local oscillator, or the use of a separate receiver, in the case of a fixed-frequency beacon.

It will be clear that the strength of the signal received by the radar set from the object on which the racon is installed is no longer determined by the radar transmitter power and the equivalent echoing area of the target but by the radiated power of the racon transmitter; reception can, therefore, be extended to the range at which the racon appears above the radar horizon of the receiving aerial no matter what the radar detection range of the object may be. As the racon transmits only in reply to a pulse from a radar set, the latter is called the *interrogating pulse*. If it is assumed at this point that the racon replies to the radar interrogating pulse instantaneously, the racon reply pulse will be superimposed on the echo of the target in which it is housed, e.g. a lightvessel or lighthouse.

In practice, although the above general principle remains true, several differences are introduced by practical expediency and to increase the efficiency of the system.

Swept-frequency racons

If a racon is to be received without the use of additional receiving equipment, its transmitter must cover the whole of the appropriate radar band. The swept-frequency racon was introduced to provide this service. Two types are currently in use, the slow sweep, in which the time to cover the 180 MHz frequency band is measured in tens of seconds, and the fast sweep, in which the band is covered in microseconds. Each will provide a distinctive signal. In the modern racons, solid-state devices are used to generate the transmission frequencies, and these are electronically tuned. It is therefore necessary to generate suitable voltages for this purpose and these are then applied to the oscillator.

The rate of frequency sweep in the slow-sweep racons is between 1·5 and 3·0 MHz per second. Since the IF bandwidth of the radar receiver may be between 5 and 20 MHz, the racon will be received for several seconds so that it should be displayed for between 2 and 7 successive revolutions of the radar aerial.

The fast-sweep racon sweeps through the frequency band in a period of around 12 microseconds. The receiver will pass the signal, therefore, for only 1 microsecond or less, and on the radar screen this would appear as a single dot. In order to provide a distinctive signal, up to twelve frequency excursions are made for each interrogation, so the racon response will appear as a series of dots.

Other broad-band racons

One of the problems of a slow-sweep racon is that the signal appears at intervals which can be up to 120 seconds. Some studies have recently been carried out on beacons that are capable of responding automatically on the frequency at which they are interrogated. This type of equipment is unlikely to be introduced into service for a considerable period of time.

In-band fixed-frequency racons

Although not in general use, some racons of this type have been installed for experimental purposes. The present problem is that a radar set has to be modified in order to receive the transmissions, though it is possible that if their use becomes more widespread, there might be some provision made for a 'beacon' mode. Current practice is to retune the radar local oscillator when the reception of the racon signal is required, and it is then theoretically displayed free of all radar signals. In practice, however, very strong radar signals can sometimes break through due to limitations in the radar IF amplifier, but their echo strength is limited.

The advantage of this type of racon is that the operator can switch in the signal as and when required. It is then displayed free of clutter and all but the strongest radar signals. It is also available continuously, as opposed to the cyclic appearance of a slow-sweep racon.

The facility for switching the signal in and out is useful since the racon signal cannot be masked by either clutter or land echoes when it is switched in, and when approached very close, no interference is experienced on the radar display, since in the normal operating condition no racon signal is received.

Identification and display

A typical display of a slow-sweep racon is shown in Plate 47 and Fig. 90(a). Due to the sweep times, the appearance of the signal is cyclic with the time of the display being considerably less than the sweep period. The overall length of the response is dependent on the operating range of the racon.

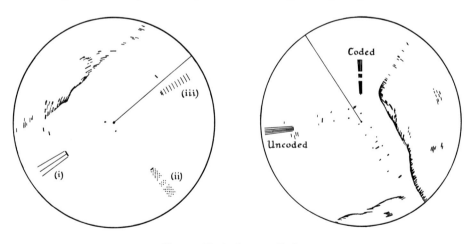

Fig. 90. *Typical racon displays.*

(a) *Slow Sweep*   (b) *Fast Sweep*

(i) *Stepped Sweep, responds only to every fourth pulse.*
(ii) *Sweep NOT synchronized to response.*
(iii) *Sweep synchronized to response.*

For medium range, up to 20 miles, it is normally equivalent to about 2 nautical miles, but for longer-range racons, the length could well be up to 4 nautical miles. A single radial line as shown is often sufficient to identify the target to which the racon is fitted, but it is possible to code the transmission, usually with a single morse-code letter, as is shown in Fig. 90(a). It is recommended that this code start with a dash, as a dot could in some cases become confused with a target echo or could be lost at shorter ranges if the swept gain control (anti-clutter) in the radar set was fully operational.

The signal from a band-edge fixed-frequency racon would be of exactly the same format as that from a slow-sweep racon, so its display is also represented by Fig. 90(a).

A typical display of a fast-sweep racon is shown in Fig. 90(b). The signal appears every revolution of the radar aerial, but because of the small size of the dots some of the signal can be lost in sea-clutter at short range. There are usually twelve dots covering a range of 10–12 nautical miles. Except by altering the number of dots in the response, it is difficult to code this form of racon.

Range measurement

In taking a range measurement, it is usual to measure it to the target echo. However, since the racon reply is initiated by the radar pulse, the racon signal is in synchronism with the PPI display, and the range observation can be made by measuring the range to the inner edge of the pulse nearest to the centre of the screen if the target echo is not visible. There is a correction to be made however: there is a delay between the reception of the interrogating pulse and the transmission of the racon signal, and this delay causes the apparent range to be greater than the actual range. The true range is given by subtracting this difference, which could be given as a fixed correction for all types of racon installation.

If such measurements are made on the racon signal, care must be taken that the swept-gain control is not operational as this could further delay the start of the racon signal at short range.

Bearing measurement

The angular width of the arc of the racon signal is determined largely by the width of the main lobe of the radar aerial. The bearing of the racon is the line from own ship's position on the screen through the middle of the racon arc. At short ranges it is possible for the racon to be operated by the power in the side-lobes of the radar beam with the result that the arc width is increased. If the radiation pattern of the radar aerial is not absolutely symmetrical a bearing error will be introduced. This effect may be slightly worse than that caused by side-lobe *echoes* because if the beacon is being triggered at all by side-lobes it will transmit at its full power. The effect will depend on the range and the degree of aerial asymmetry. This error can be very largely removed by reducing the receiver gain until the arc is as narrow as possible although painting consistently.

Spurious responses may appear on incorrect bearings due to triggering by reflections from objects in the vicinity of the racon site. Similar results might occur due to reflection from obstructions on board the ship (see False Echoes in Chapter 7). Reduction of receiver gain will usually remove the former from the screen.

A further source of error in bearing measurements can occur when using a band-edge frequency beacon. The modern merchant marine radar set usually uses an end-fed slotted waveguide aerial in which the direction of the main lobe is influenced to a certain extent by the frequency. As this changes, the angle which the main lobe makes with the aerial changes by a small amount known as the squint angle. If the racon and radar frequencies differ, a small error due to this squint angle will be introduced. In practice this error will be limited to about 1°.

IN-BAND RAMARKS

This type of radar beacon is very little used today, apart from some rather specialized applications. It is a continuously transmitting beacon which will display a radial line of bearing on the PPI. In some ways it is analogous to the rotating coil M/F radio direction finder, which has been in use for many years, although of course the radar aerial is much more directional than the D/F coil. In the cathode-ray tube direction finder, the bearing is displayed as a radial line on the tube. In the radar case, if the ramark transmits continuously on the frequency to which the radar set is tuned, then when it is within the beam-width limits of the radar aerial, the signal received will appear on the screen as a bright radial line or a narrow sector, on the bearing of the beacon.

One of the navigational requirements is a characteristic signal which will enable one ramark to be distinguished from another. There are two possible methods of achieving this 'coding', first by

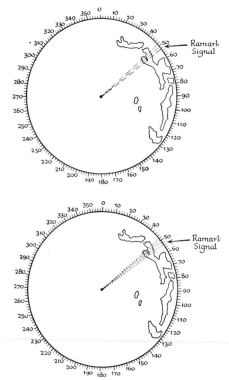

Fig. 91. Ramark coding

having the bearing line on the PPI broken up into dots or dashes, Fig. 91, and secondly by arranging that the signal is displayed only on some of the revolutions of the radar aerial.

Coding by either method may be effected by appropriately selecting the rate at which the frequency is varied. As with a swept-frequency racon, signals will only be received when the ramark frequency is within the IF pass band of the radar receiver. For fast sweeps this period will be very short, but with slower sweeps it will be extended, giving longer lines on the PPI display.

If, for instance, the sweep rate is high, e.g. 100,000 times per second, the signal will appear as dots on each scan, a number of scans being displayed within the beam-width of the aerial. An example of this form of signal is shown in Plate 46, which depicts two ramarks adjusted to display dots. A slower rate of sweep will display dashes, and experiments have shown that the great difference of sweep rate necessary to distinguish with certainty between dots and dashes probably precludes using more than two identity codes.

There are also two methods of time-coding. One is to use a sweep speed similar to that used in the slow-speed racon so that the ramark is displayed periodically. The second is to use the high rate of sweep but to switch it on and off in a regular manner.

COMPARISON OF RAMARKS AND RACONS

The in-band ramark provides a constant and virtually omni-directional transmission which can be received by any ship equipped with radar operating in that band. The characteristics of the transmission, apart from time coding, permit the use of two identity codes (dot and dash) and the rate of supplying information is sufficient for pilotage. If time coding is used additionally, the total number of codes available for use as a landfall aid will be about six. The signal is superimposed directly on the radar display in the form of a bearing line comprising dots or dashes.

The slow-sweep method of time coding would be unsatisfactory for use with sets which have a narrow i.f. band-width (about 4 MHz).

The racon has the advantage over the ramark that it provides all the information required for a fix, i.e. range in addition to bearing and identification of its location. To receive this information from band-edge and cross-band racons, the radar set needs to include special receiver facilities.

Any device in which the signal can be superimposed on the radar picture has an advantage in pilotage or close waters. In this connection there is a possibility of using ramarks as leading lines without losing the use of the radar.

The accuracy of bearings taken from ramarks or racons is essentially determined by the characteristics of the radar set. If care is taken to ensure that the PPI is accurately centred, parallax avoided, and the receiver gain adjusted to give the narrowest bearing line, the bearing accuracy should be about 1° when the radar set is used as cathode-ray tube direction finder. It should be remembered, however, that the primary function of a ramark is to *identify* a point on a charted coast.

Radar beacon interference

The use of in-band radar beacons of both the ramark and racon type inevitably brings with it the probability of increasing the extent of interference with the normal radar picture. This interference can arise from a number of sources but by far the most serious is that of triggering the racon from the radar aerial side lobes. Fortunately this form of interference is only experienced at close range and development in aerial design has further improved the situation. Side lobe reception can also be a problem in the use of ramarks.

During the development phase of radar beacons a considerable effort was devoted to means of minimizing this interference by suitable design of the beacon and by careful selection of its transmission characteristics. A great deal can be achieved by matching the transmitter power output to the operational range and in the case of a racon, by designing its receiver sensitivity so that it only triggers the racon when it is interrogated by ships within the operational range. Too high a sensitivity in the racon receiver will increase the interference due to side lobe triggering as well as to ships outside the designed operational range triggering the racon unnecessarily. Careful design of the radar beacon can only reduce the degree of interference; there will always remain the prospect of experiencing interference at close range; further steps are therefore necessary to enable the navigator to remove any interference from his radar screen.

The identifying paint on the PPI can be made of any convenient length by selecting an appropriate pulse length for the radar beacon. A relatively long pulse, compared with the normal pulse length of a radar set, makes it possible for the radar observer to remove the radar beacon paint from the PPI by operating his differentiator (rain-clutter) control.

Interference may also be experienced due to reflections from objects in the vicinity of the radar beacon or from masts, funnels &c. on the ship. It will usually take the form of a synchronized but spurious response on an incorrect bearing. Careful siting of both the radar aerial and radar beacon will assist, but in extreme cases it may be necessary to employ radar absorbent material on the object which causes the spurious reflection.

During periods when a large number of ships are interrogating a racon, random interference may be experienced due to the reception of unsynchronized racon transmissions. Once again, the use of the differentiator will reduce or eliminate this effect, if the racon transmissions are of the long pulse type.

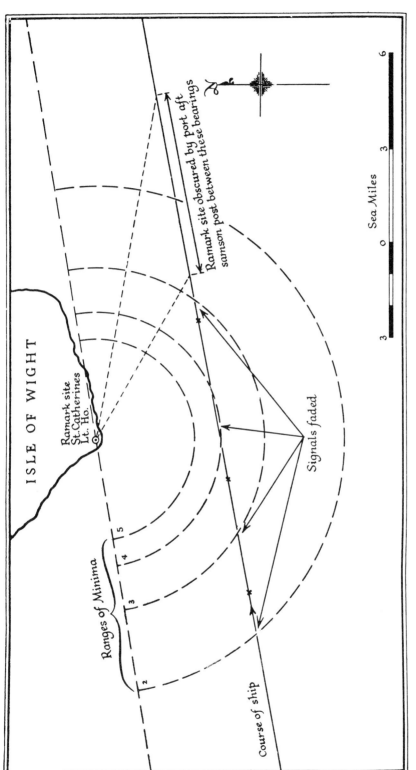

Fig. 92. Periodic loss of ramark signals due to interference minima.

The use of ramarks and racons

The maximum reception range of ramarks and racons in the 3-cm. band is that at which they appear *above* the radar horizon of the radar aerial; it is unlikely to be regarded as a limiting factor, though it will of course be affected by precipitation and by the prevailing conditions of propagation (e.g. by super- and sub-refraction). As in the case of radar reflectors, reception will be affected also by the interference minima set up by the interaction of direct and reflected rays.

Fig. 92, which is typical of these interference minima, illustrates the loss of ramark signals actually experienced by a ship passing an experimental ramark at St. Catherine's point.

Time coding may also be prejudiced by these minima and is liable to ambiguity unless the transmitter power is made much greater than would otherwise be necessary.

These beacons are likely to give the best signal on the PPI when the long-range scale is in use, even when at no great distance from them. The sea-clutter control will affect the beacon signal when the i.f. amplifier of the radar set is part of the receiving circuit, and it may require adjustment. When using such beacons for navigation, the north-upward display will usually be found more convenient.

NEW RADAR DISPLAYS AND THEIR EFFECTS ON RADAR BEACONS

Recent developments in civil marine radars have produced new forms of displays. Some of these use video processors to produce target echoes of greater clarity and to reduce the effects of clutter. These processors can suppress the signals from some in-band racons so that they are not displayed. Integrating type displays are also available and because of the cyclic nature of the slow-sweep racon, the response may not always be as satisfactory on this type of display as in the standard radar presentation. However, in both cases a standard display is also available, and it is this which would be employed if use is to be made of the racon signals.

12

SHORE-BASED RADAR

In the early days of commercial marine radar during the immediate post-World-War-II period, the radar-fitted ship was an unusual phenomenon. Even by 1950, of the ships arriving at any major port the percentage that was equipped with radar had barely reached double figures. Although these ships could use their radars to good effect at all times in open or in coastal waters, their advantage was not so marked in the congested waters of a large fog-bound port. In such a port, the many ships arriving without radar were forced to anchor in the approaches, maybe without knowledge of their precise position and often obstructing the channel. Berths and anchorages rapidly became congested and in all these circumstances the master of a vessel, even if fitted with radar, would generally have considered it imprudent to commit himself to proceed either inbound or outbound, as he would be unaware of what hazard he might encounter once under way, and possibly at an awkward place in the channel. The Port Authorities were unable to provide any useful information as they too were lacking any accurate and detailed knowledge of the situation. When the fog eventually lifted there were further delays due to the congestion and overloading of the resources of the port—pilots, tugs, berthing space, cargo handling capacity etc.—and it was not uncommon for it to take upwards of one week to get the port operating more or less normally again.

At about this time a few far-seeing Port Authorities considered that the establishment of a carefully sited Shore Radar Station of suitably high performance could provide them with the missing information about the position and movement of all vessels in the main approaches, channels and anchorages. This vital information could be passed on to masters and pilots of ships involved, if suitable radio telephone facilities were made available, thereby making it possible to maintain and *regulate* movement of shipping in fog. The Mersey Docks and Harbour Board were leaders in this field and established at the port of Liverpool in 1948 the first purpose-built Harbour Surveillance Radar and Communication Station. Now, in 1977, it is estimated that upwards of 300 shore-based marine surveillance radars of various degrees of complexity are in operation round the world.

The functions of a harbour surveillance radar system

A suitable radar system, which, in a large and complex port, may incorporate several radar equipments, shows the positions of all vessels and many of the navigation marks such as buoys and lightfloats. In conjunction with other information obtained by radio, telephone, telex or documents from various sources ranging from owners, shipping agents to ships themselves, this enables the Port Operations Organization to have available at all times an up-to-date total picture of the traffic situation in the port and its approaches. This information is made use of in the following main activities:

(1) Maintenance of safety in all traffic movements—this is of paramount importance.
(2) Expeditious regulation of traffic movements, by the interchange of information with ships.
(3) Provision, on request, of detailed navigational information to assist the master/pilot of a vessel under way.
(4) Monitoring and assisting the co-ordination of Search and Rescue (S.A.R.) and pollution control operations.
(5) Providing data for statistical and operational traffic analysis.
(6) As a 'housekeeping' aid, monitoring the positions of dredgers, spoil-dumping hoppers, buoys and various Authority patrol vessels.

In all these activities, first-rate communications are an essential prerequisite for success. It is significant that when the first Harbour Surveillance system was installed at Liverpool in 1948, the problem of immediate communication with the pilots of vessels was recognized and solved by the use of portable VHF/FM radiotelephones taken aboard by the pilot, one of the first uses of the then newly-allocated Maritime VHF band. Today, virtually all ships are fitted with VHF radiotelephones and Port Radar/Radio Operations rooms are equipped with very comprehensive multi-channel VHF and MF systems. On the shore side, extensive direct-line and switched telephone networks are commonplace, often supported by teleprinter and telex facilities.

Although the above paragraphs have referred exclusively to Port or Harbour operations, almost parallel requirements and activities exist in the surveillance of certain stretches of estuarial and coastal water. This is especially so where there are either heavy concentrations of maritime through traffic, maybe completely international in character, such as in the Dover Strait or Malacca Strait, or where there is some special pollution or other hazard, such as a considerable oil-tanker traffic. In North America the term Vessel Traffic Management System is used to cover all applications and seems an apt one.

General technical aspects of shore-based surveillance radar systems

The detailed technical requirements are, of course, related to the operational requirements and to the physical features of the area to be covered. In many cases the maximum range and both range and bearing discrimination requirements can be met adequately by standard marine radar units. The use of these standard units, giving reliable detection of all craft over say 30 ft. length to a range of 5-6 miles with an aerial beamwidth of $\frac{3}{4}°$, offers the economic and logistic advantages associated with quantity production. Examples of the many installations of this nature are those at Dover, Folkestone, Scheveningen, Adelaide, Gothenburg, Los Angeles and over 100 others worldwide, many used directly by pilots.

At the other end of the scale are the port and coastal water areas whose surveillance require either extremely good discrimination in bearing and/or range or very long-range detection capability; often both are required. Usually, because of the extensive area to be covered, it is necessary to provide several PPI display units with different degrees of off-centring and maybe range-scale factors, all fed from the same radar aerial and transmitter-receiver. A typical example of this type of installation at Liverpool is described in detail later.

In many cases, the topography of the area makes it impossible to select one radar site from which adequate coverage can be obtained. Typical examples of this restriction occur in the approaches to estuarial ports such as Teesport (described later), London, Hamburg and Rotterdam, and in large area surveillance as in the Dover Strait (also described later), the approaches to the Vancouver area and in several places on the Canadian Eastern Seaboard. Very frequently the optimum sites for the radars are remote and inaccessible and it is both operationally and administratively unacceptable to establish a multitude of display and operations rooms at these positions. In such cases one conveniently sited operations centre is established and the radar data is brought in by microwave links from the remote sites to the displays in the centre. The radar sites are totally unmanned and are remotely controlled from the operations room.

In all but the simplest systems, all the units (except the scanner itself) are duplicated and interswitched to provide continuity of operational facilities in the event of component failure. This duplication is extended to the microwave link and remote-control systems and frequently to the electrical power supply, where automatic-start diesel generators are provided, capable of restoring full operation of the system within 10 seconds of main supply failure.

In the operations room the display arrangements are invariably of the PPI type but frequently incorporate special features to meet the local operational requirements. In most cases some degree of picture shift is required, by offsetting the radar trace origin from the tube centre, to increase the tube area available for the region of interest. In extreme cases as much as 5-radii off-centring is required in order to display on a large enough scale a particularly sensitive area some distance from the radar site; Liverpool provides a classic example.

The provision of means for rapid and accurate measurement of echo position is required in many cases. Very frequently it is of much more value to the master of the ship if positions can be measured and passed to him in terms of bearings and ranges from some convenient navigational feature (such as a buoy or light) rather than from the shore radar site, the position of which may have no navigational significance. This is achieved conveniently by the provision of an 'interscan' line which can be adjusted in length and direction and whose origin can be moved to coincide with any echo on the face of the tube.

As the name implies, this 'interscan' line is produced on the face of the tube during the intervals between normal 'mainscans' when the cathode-ray 'spot' would otherwise be idle and waiting for the next radar trigger pulse. As it is generated and displayed very frequently, once per radar pulse is possible although a rate of 40–50 per second is more usual, the line is continuously visible. It is therefore much more rapid in use than the simpler variable range marker and adjustable bearing flash marker, both of which are painted by the main trace and therefore relate to any particular echo only momentarily, once in each rotation.

The accuracy of even an 'interscan'-type of measurement system is limited to about 1 per cent in range and 0·5° in bearing and in some circumstances this is inadequate. Modern digital integrated circuit technology has made it possible to generate very accurately defined pulses which may be mixed in with the normal video signals from the radar receiver. Patterns of these are used to mark channels, leading lines, buoy and light-vessel positions and the like, with an accuracy of a few metres. Plate 51 shows the mandatory IMCO traffic-separation lines in the Dover Strait, as presented by this technique. In a few other installations, digital computers are used for controlling accurate digital markers on the display or for automatically extracting position directly from the radar video.

With a conventional type of PPI display using a long afterglow phosphor, the picture cannot be viewed satisfactorily in normal ambient lighting. As the use of light-excluding viewing hoods on the multiple displays in a large operations room is not practical, the room has to be designed for working in a low ambient light level at all times. If the operations room staff also have to be able to maintain visual look-out, the windows have to be fitted with variable-density light screens which have to be removed at night. Alternatively a separate operations room is set up, this being totally enclosed and artificially illuminated at low level at all times.

These conditions can prove oppressive for the staff and special high-brightness displays, capable of being viewed in normal daylight, are being specified more and more often. These may use either the rapid photographic/projection technique as currently fitted at Teesport or, increasingly, conversion of the radar scan data to television signal format by means of either analogue or direct digital-scan converters. The scan conversion systems present the final picture on high-quality TV monitor displays which can be viewed in all but direct sunlight and have substantial afterglow. They have the further advantage that extra monitor repeater displays are relatively inexpensive and simple to add. Very roughly, the radar picture on a small, high-definition CRT is photographed by a TV camera and displayed on a TV-type tube, thus converting the rotary radar scan to a TV raster. Such systems are found at Le Havre, Medway, almost all the Canadian installations and in the replacement systems to be installed at Teesport in 1978.

LIVERPOOL SHORE RADAR STATION

The first installation of shore radar equipment specially designed for harbour surveillance was put into operation at Liverpool in 1948. It was completely replaced by more modern equipment in 1959 and again in 1973. The two earlier installations were at the seaward corner of Gladstone Dock on the NE bank of the mouth of the river. Following the reclamation of an area immediately downstream of the radar and the subsequent building of the large new Seaforth Dock and Container Terminal complex in this area, it was decided to resite the radar and operations room at the seaward (NW) corner of this reclaimed area at the time of re-equipment in 1973 (Plate 50).

The channel at Liverpool extends some 13 n. miles from the mouth of the River Mersey out to the Bar Light-vessel and is marked for most of its length by large buoys at about 5-cable intervals on either hand. The channel has a number of bends and in places narrows to only 1540 ft. (470 m.) The long range and the very high discrimination required in this application demand the use of a large aerial to give the necessary system gain and the narrow beam-width, as it is not practical to site a remote radar head anywhere closer to the seaward end of the channel. The new site gives an uninterrupted view over the whole of Liverpool Bay and up the river as far as Eastham Locks at the entrance to the Manchester Ship Canal.

A hog-horn fed parabolic reflector aerial, with a horizontal aperture of 25 ft. and a vertical aperture of 2 ft. provides beam-widths to the half-power points of 0·3° horizontally by 4° vertically. This aerial, operating in the 3 cm. band, is rotated at 20 r.p.m. by a 10-HP motor, and is mounted approximately 100 ft. above high-water level on top of a reinforced concrete tower. Below the aerial is the operations floor with visual lookout, the seven PPI displays being installed in a completely light-locked radar operations room behind the visual lookout area.

The radar equipment, also housed on the same floor as the operations room, has duplicated 20 kW transmitter-receivers of marine type operating on 9170 MHz with switchable pulselengths of 0.05 and 0·2 microseconds, at a constant PRF of 2000 Hz. The receivers have logarithmic response (Plate 49).

The seven 16-inch-diameter PPI displays are fitted in a curved console which also houses complete R/T and landline communication operating facilities. One of the displays (Plate 48) is normally set to show an off-centred general view of the whole of the channel and approaches in Liverpool Bay to a range of approximately 20 n. miles, while the other six show overlapping sections of the approach channel and river at enlarged scales (approximately 3 in. or 5 in. to 1 n. mile) and with varying degrees of off-centring, in one case amounting to more than 4 radii. These range scales and off-centring are preset and each display has an 'area selector' switch instead of the normal range switch found on shipborne radars.

Any given area can be selected on more than one display to provide continuity of cover in the event of breakdown or during maintenance activities or any special operational circumstance where two operators may require coverage of the same area.

The displays are fitted with a Computer Assisted Measurement System which provides facilities for rapid and accurate measurement of ship's positions relative to any of 24 reference points in each display coverage area. The reference points may be preset e.g. on to buoys, fixed marks, etc., and some may also be set by the operator on to, for example, a temporary navigation mark, a dredger, an anchored vessel or a wreck. The system also provides facilities for measurement of Track Made Good, and Closest Point of Approach of a vessel to any of the reference points; a dotted line ahead of a vessel predicting its track can also be shown. Interscan markers for range and bearing measurement are provided as a stand-by to the computer-aided system. Reflection plotters and mechanical cursor grids are provided in case all else fails.

In addition to a multi-channel voice recorder system, a video tape recorder is installed which permits recording of the entire radar video signal for later display either for training purposes or for subsequent analysis of any particular shipping movement.

Methods of conveying radar information

To enable vessels navigating in the river and sea channels to make full use of the information available from the radar station the ship must be in direct VHF radio communication with the station. It is mandatory under Liverpool Port regulations for all vessels of more than 50 tons gross to carry VHF R/T equipment in proper working order. In the event of a vessel not having such equipment in working order, the Port Authority will hire out portable VHF equipment for use when the vessel is navigating within the Port. Stocks of these portable VHF radios are kept at the Pilotage Office in Liverpool, aboard the Liverpool Bar Pilot Boat, at the Point Lynas Pilot Station which is 33 miles west of the Bar and at the Port Radar Station itself. Six channels of VHF radio are fitted at the Port Radar Station in addition to the calling and safety channel and five of these channels plus time injections are continuously recorded. Five additional channels are shared between the various dock entrances for detailed docking instructions when vessels have entered the river.

The positional information provided by the displays may be transferred to a chart or transmitted to a ship using any of the following methods:

(i) *Bearing and distance from the Radar Station.* This provides an accurate position but is somewhat cumbersome unless the target is close to the station, as it requires plotting on a chart.
(ii) *Giving the position as a true bearing and distance from the nearest known object.* This might be a lightvessel, lightfloat, buoy, pier

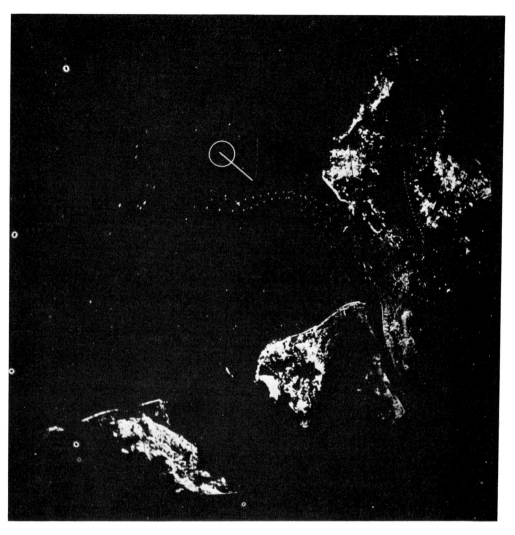

Plate 48. *Liverpool Bay, overall PPI picture, showing channel marking buoys, etc.*

Plate 49. Liverpool Radar Operations Room. (Only 5 of the 7 displays visible.)

Plate 50. Liverpool Tower.

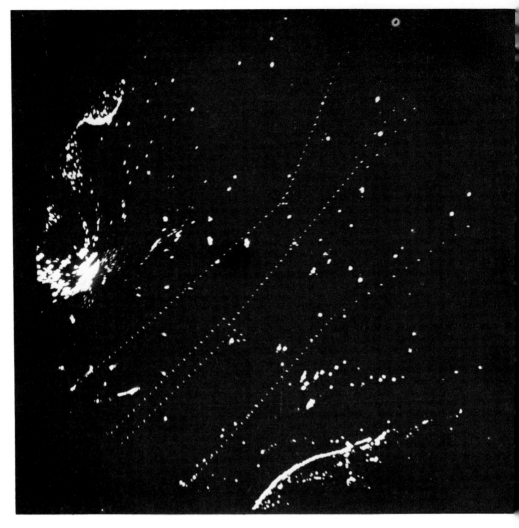

Plate 51. *Dover Strait from St. Margaret's Bay.*

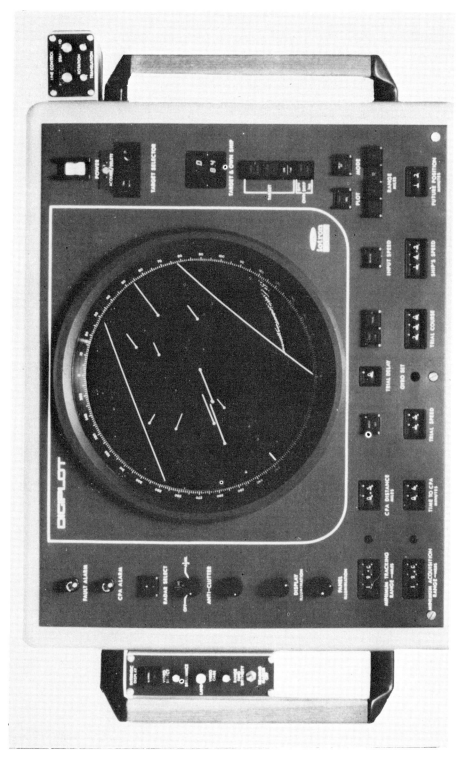

Plate 52. Digiplot.

or jetty and the usual practice is to use the next object ahead of the vessel. This method has been found to be the most practical way of conveying information.

(iii) *Giving an approximate position relative to the centre line of the channel or to the line of buoys.* In practice this method is used, in addition to method (ii), in the form of a running commentary in the intervals between the transmissions of more accurate positions. Vessels find the approximate information of great value when under way in the channels.

Operational procedure

The equipment is operated continuously.

A disposition statement is broadcast for the benefit of all vessels, at 3 hours before each high water and repeated with any amendments at 2 hours before high water. Information regarding the state of the weather, visibility and any urgent local navigational warnings is broadcast at the end of each disposition statement. Regular routine checks are made of the positions of all the floating seamarks. A tide gauge repeater giving the tidal height at the Gladstone River Entrance is fitted and this information is available on request.

All this information is of considerable use in clear weather since the state of the river and approaches will not be evident to a ship from a distance, even if her radar is in use.

If a vessel is using her own radar when navigating the channel in low visibility, the long-range information given by the shore radar station may permit her to keep her own set on the shortest range scale and thus in continuous use for pilotage.

The experience gained since the station was opened has shown the value of the running commentary, which the ship acknowledges at intervals of about five minutes; a more frequent acknowledgement seriously reduces the amount of information which can be passed.

The problem of the identification of individual vessels requiring assistance has not proved difficult. If a vessel is in the approaches and has plenty of sea room, identification can be made by requesting the pilot to alter to a pre-determined course at an agreed time. A vessel which has entered the channel can identify herself at a particular buoy with subsequent checks as she proceeds up the channel. Once a vessel has been identified, her track is plotted continuously on the reflection plotter.

It is essential for the staff of such a station to be competent not only to operate the radar equipment but to appreciate the significance of the information displayed and transmit it to masters and pilots in the form most readily understood. This object will be achieved more rapidly if the operators have had sea experience. All the Duty Officers at Liverpool Port Radar Station possess a foreign-going Master's Certificate.

TEESPORT RADAR SYSTEM

The present installation at Teesport is intended to provide precision radar coverage of the entire estuary, the approaches and the navigable part of the Tees. The radar used is mainly the Kelvin-Hughes PRP (Photographic Radar Plot), employing two 15 ft. slotted waveguide aerials mounted on 100 ft. towers. One of these is sited at the Harbour Master's Office, which houses the control room, alongside the Tees dock. The other is almost three miles away at South Gare, the southern promontory in the Tees estuary. The radar data from the remote scanner is relayed back to the control room and the control information sent in the reverse direction by means of a microwave radio link.

The control room equipment consists of three 24 inch PRP displays and one 16-inch conventional direct viewing radar. Two of the PRP's are integrated to give a continuous radar map of the whole port area; the third covers a proposed anchorage area and the approaches to Hartlepools. The conventional display is used for general surveillance or when traffic is light.

After more than 10 years' operation, this equipment is due to be replaced in 1978. The new system will consist of 4 radars at 3 sites. The South Gare site will have an X-band (3 cm.) radar with an 18 ft. scanner and an S-band (10 cm.) radar with a 12 ft. scanner. Both will be microwave-linked to the Operations Room at the Harbour Master's Office. The operator will be able to choose between the higher discrimination of the X-band and the improved clutter suppression of the S-band radar. A new X-band radar also linked to the Operations Room will be established at Dabholm Beck, situated between the two original sites, to provide coverage in areas which have been developed and others which have been obscured since the earlier installation. The Harbour Master's Office accommodates the Control Room and the third radar site, X-band radar with 15 ft. scanner.

The new system will incorporate five 17-in. daylight viewing TV-format displays operating from two scan convertors, as described earlier in this chapter. A conventional PPI display will also be included. Video tape recorders will be incorporated, recording the TV-format signals rather than the raw radar signals which are recorded at Liverpool.

CHANNEL NAVIGATION INFORMATION SERVICE

H.M. Coastguard runs this service from its Operations Centre at St. Margarets Bay, three miles NE of Dover. It uses two radars, one at St. Margarets Bay and the other at Dungeness, linked to the Operations Centre by microwave circuit. These provide surveillance of the Dover Strait Traffic Separation Scheme (Plate 51) from the meridian of Greenwich to the vicinity of the West Hinder Lightvessel. The French authorities operate a closely co-ordinated service from CROSSMA, Gris Nez. The aim of the joint service is to assist safety of navigation through the

Strait by providing a service comprising up-to-date bulletins of strategical and tactical information.

Bulletins are broadcast at 30-minute intervals by the British on VHF channel 10 at 10 and 40 minutes past each hour, and by the French on VHF channel 11 at 20 and 50 minutes past each hour, followed by a translation in English. If the visibility falls below two miles, the bulletins are broadcast at 15-minute intervals.

Bulletins contain information on:

(a) Visibility, way-mark faults, ships broken down, floating mines and any other background intelligence.

(b) Positions, course and speeds of ships which appear to be acting in contravention of the Regulations for Preventing Collisions at Sea (1972).

Masters who are unable to conform with those Regulations are invited to communicate with Dover Strait Coastguard (VHF Channel 10) or CROSSMA Gris Nez (Channel 11). Both centres operate sea and air patrols to identify vessels which appear to be contravening the Collision Regulations and they report to their respective Administrations the facts of each case for potential legal follow-up action.

The Equipment

An initial installation in 1970 at H.M. Coastguard Station at St. Margaret's Bay, overlooking Dover Strait, of a conventional marine radar proved the requirement for extensive radar coverage of the Strait to monitor the degree of compliance of shipping with the IMCO Traffic Separation Regulations.

In 1976 this radar was replaced by one fitted with a 25 ft. aerial and duplicated transmitter-receivers. These units are identical to those at Liverpool except that the aerial rotation rate is 12·5 r.p.m. and the pulse lengths available are 0·2 and 1 microsecond at a PRF of 1000 Hz. A similar radar was also installed on the roof of the Dungeness Nuclear Power Station approximately 22 n. miles SW of St. Margaret's Bay, the radar data being fed by duplicated microwave radio link to Swingate (where a suitable tall tower already existed) and thence by duplicated underground cable to the Operations Room at St. Margaret's Bay.

In the Operations Room three 16-inch fixed-coil PPI displays are provided, each with 4 range scales of 4, 8, 16, and 32 miles radius and full radius off-centring. Each display can be switched independently to show either the Dungeness or St Margaret's radar data. Normally, two are fed from Dungeness and one from St. Margaret's, each on 16-mile range-scale and off-centred so that the three together cover about 70 n. miles of the Channel from S. of the Royal Sovereign Light-Tower through the Strait itself to N. of the Falls Lightvessel. The high effective power of these radars, coupled with their elevated sites (250 ft. above sea level at Dungeness and 300 ft. at St. Margaret's), provides reliable detection of 30 ft. wooden trawlers to a range in excess of 18 n. miles.

Each display is equipped with an off-settable interscan bearing and range line and both radar channels are fitted with Deccaspot generators as described earlier, which are programmed to show the mandatory IMCO separation lines and zones to an accuracy of a few tens of metres.

It is planned to move the operations room to a new site above Dover Harbour in 1980 and to extend the display facilities at that time.

13

THE IMPORTANCE OF RECORDING EXPERIENCE

It is not always appreciated that keeping a record of the performance of a radar set in some brief but readily comprehensible form is a most important aid to the efficient use of the equipment. It is fully realized by the authors of this book that modern seagoing officers have a wide variety of duties many of which entail paper work, but the importance of a radar operating log is such that it has been thought worth while to devote a chapter to emphasizing the reasons for keeping one and suggesting a methodical way of doing so. Although a well-kept log may contain a mine of information of value to those responsible for providing and designing radar, it would be a great mistake to assume that this is the major reason for recording the results obtained. On the contrary, the user who records his experience will reap the greater benefit from possession of the log.

Two factors which affect the operating value of radar are the state of efficiency of the set itself and the meteorological conditions. The operating efficiency of the observer is a third factor which may be of more importance than the other two. When circumstances are such that other aids to navigation are not available or are reduced in value, it becomes of the greatest importance to those responsible for the safety of the ship to know what may be expected of their radar and to be able to judge from its current performance if the set is at full efficiency.

The performance monitor should relieve anxiety on the latter question, but it will not tell what may be expected *in terms of targets and ranges* and its evidence will be no guide to the meteorological conditions and their effects in those terms, since it is substantially unaffected by them. The navigator will be able to make a practical and confident appreciation of the overall conditions only by comparing the response indicated by the echoes which now appear on his PPI with his previous experience against the same or similar targets. If, however, this confidence is to be soundly based, he must have some yardstick against which to draw comparisons. The best criterion, of course, will be the results obtained by his particular radar set when at full efficiency and in the normal weather conditions for the area he is in. For this, it will no doubt be agreed, he should not have to depend upon a feat of memory when it is possible for him to have at his elbow a record of performance on the widest possible variety of targets.

There is also the question of recognizing particular land echoes when making landfall or passing within extreme range of land on passage. It has been observed in Chapter 8 that this is not an art which can be mastered on first acquaintance and that observations made on previous occasions in clear weather may prove to be a boon in low visibility. This dividend will be paid in far greater measure if a pictorial record of the radar appearance of notable landfalls is kept and referred to.

Lastly, there is the case of the 'new hand', the relief navigator and even the relieving officer of the watch. In the case of officers new to the ship or route, or perhaps new to radar observation, knowledge of what to expect of the radar will set them well on the way to efficiency in observing. But this cannot be passed on to them in a five-minute turnover; something is needed for them to refer to at any time during their watches or between them. Similarly a turnover to any officer relieving the watch can be more explicit if the yardstick is available in terms of accumulated and recorded experience.

The habit of recording the behaviour of the radar set inevitably makes the observer familiar with any idiosyncrasies the set may exhibit and removes any aura of mystery which may surround it. It also encourages the observer to be critical of the performance of his set and thus enables him to provide the maintainer or service engineer with information that may enable him to prevent breakdowns and to clear defects quickly when they do arise. A well-kept radar log will be of considerable help to the maintainer when the observer is not present to describe the circumstances leading to a breakdown.

An operational log can be of great help to those who are concerned with the development of radar, whether in manufacturing equipment or in developing ancillary devices such as corner reflectors, &c., or in any other way in which information from sea provides the data which can guide further progress. It is also, of course, of interest to those responsible for providing radar in their ships since the information it contains will give a useful indication of whether the expenditure is justifying itself in measurable terms.

Thus it can be seen that the effort involved in keeping a log will repay the seagoing user both directly, in the valuable information it can yield, and indirectly in furthering the improvement of radar facilities.

THE FORM OF THE LOG

To meet the requirements of all parties, the log must be something more than a list of times ON and OFF with a few general remarks about performance. A useful log might conveniently be divided into two main parts, a Log of Targets and the Operational Log.

Log of targets

The log of targets will be kept mainly for the benefit of the user himself and is intended only for entries of lasting interest.

It cannot be emphasized too strongly that the only entries of permanent use are those which include full details of the circumstances at the times of observations. Earlier chapters have described the large number of factors which may affect the performance of radar against various targets. As a rule it will be seldom that complete evidence can be recorded, but full entries on a few occasions will be worth very much more than pages of entries which tell only half the story.

To form a basis for comparison between current performance and that which should be obtainable it is necessary to know as much as possible about the latter. As this will vary from ship to ship and set to set, it can be obtained only by observations made in the particular ship. The object of this part of the log is to record ranges obtained on as many types of target as possible and to select those which appear to be the best for each type as the criteria for comparison with subsequent results. In making the selection, observations made when anomalous propagation is suspected should be excluded, unless the ship is in an area where super-refraction is common.

The following principles and considerations may be found helpful when making entries in the log.

Identification of targets. Whenever possible, targets which can be fully identified or described should be chosen for recording. In the case of ship targets the record will be of much greater value if the circumstances which affect radar response are included, e.g. size, type, aspect.

State of weather and sea. The weather and sea conditions are always of interest and the former are especially important when, for example, super-refraction may occur or when there is precipitation. Generally speaking there is a lack of information on the effects of weather on centimetre-wave radar. Chapter 5 has dealt broadly with the effects of anomalous propagation and precipitation but there is little practical knowledge at the present time regarding the limitations imposed on radar operational performance by sub- and super-refraction and by attenuation in rain, snow and sandstorms, &c. Thus, besides giving more weight to observations on targets, accurate details of the weather conditions obtaining will be a valuable addition to the data on meteorological effects.

With precipitation in particular, it will often be possible to obtain information of permanent value to all who are interested in the subject. Besides recording the nature of echoes from rain, hail and snow, it will be most useful if ranges or descriptions can be given of the echoes of unobscured targets compared with those when the same targets are within or beyond a precipitation area. On such occasions, information concerning the character and degree of precipitation is very important, though naturally it may not be obtainable.

Sea conditions are important when recording information concerning small targets such as buoys, boats and sandbanks.

The performance monitor. The performance monitor will be a general guide to set performance, and it will be an advantage to include its reading when logging observations. Its maximum reading should be included under Permanent Information (see below).

Detection ranges. As has been explained, one object of recording results is to enable the observer to provide himself with a criterion of what his radar equipment should be able to achieve against each particular type of target. It is probable that no very accurate estimation can be made, and it is not necessary that one should be. It will be clear that, with some types of target, the response which can be expected will be conditioned by factors other than those of set efficiency and meteorological conditions. Thus, the response of ships is affected by their aspect, that of small targets by the state of the sea and that of islands, wrecks and sandbanks by the state of the tide. Even if great pains are taken to specify all such obviously relevant particulars, it will not be possible to forecast with accuracy the effect of slight changes in them on subsequent occasions. In the case of ship targets, of course, the aspect cannot be forecast. An average of results obtained in normal meteorological conditions when the radar is known to be working efficiently will be quite adequate for the purpose in mind.

When comparing reports of the same target from different ships it is most helpful if there is uniformity in the terms used. It is suggested that the average of a number of ranges at which a particular target or type of target is first detected should be called the *average detection range*. It is desirable to specify a little more closely what is meant by this term. The *perceptibility* of an echo depends essentially on the consistency with which it will paint. The range at which a target is seen for the first time as a fleeting echo, painting perhaps once in ten revolutions of the scanner, may differ quite considerably from that at which it will paint on every revolution. In order to give a definite significance to 'detection range' it is necessary to choose an arbitrary degree of perceptibility. An echo which paints five times in ten revolutions of the scanner should be 'easily perceptible' and this therefore is a practical degree to employ.

The range at which a particular target becomes easily perceptible on any given day may be greater or less than has been estimated from previous experience as the average detection range. It is nevertheless well worth recording this in the log. A term in use for such a range is *first detection range*.

While these principles can readily be followed for ship targets and for isolated and prominent land features it is not so easy to decide upon the detection range of a coastline. Although isolated echoes of high land may paint at long ranges it is usually difficult to identify precisely the source of the echo to which the term should apply because, by the time the coastal echoes have assumed a recognizable shape, the original echoes will have merged into the general picture. Also it will be remembered that slight changes in the angle of view may shift considerably the points from which the echoes are being received. Nevertheless continued recording of the ranges at which particular portions of coastal areas become easily perceptible will form an accumulation of evidence regarding important landfalls which may be of the greatest value in thick weather.

A more definite criterion, which however will not reduce the value of the evidence just mentioned, may be the range at which a particular coastline becomes *identifiable*. Admittedly this also may vary considerably with aspect and it will certainly depend greatly on the experience of the observer. Taking all these factors into account, however, the range at which identification of the coast becomes certain is of extreme importance to the navigator and is a good guide to the performance standard of the radar set. There is no doubt that it is worth recording and that a clear distinction should be drawn between it and the range at which easily perceptible but unidentifiable echoes are first received.

The Radio Advisory Service, which was operated by the Chamber of Shipping, designed a Log of Targets, which is used by several shipping lines and may be obtained from Witherby & Co. Ltd., 32–36 Aylesbury Street, London EC1 0ET.

How to make entries in the log. If an entry in the log is to include details of the prevailing conditions, keeping a log on a blank piece of paper would indeed be a laborious affair. The use of sheets with headed columns will simplify the work a great deal and ensure that all relevant details are included.

To reduce the amount of writing still further, it is helpful to employ special symbols and abbreviations. For example, if a range is a first detection range (as defined above), it could be ringed round in ink. If then, or later, it is decided to regard it as an average detection range it could be ringed twice, so that, when referring back through the log for comparable observations these results will stand out. A range which has the qualifications of a first detection range but is obtained on a receding target might be logged as for instance, 12 R, ringed. No doubt some navigators will like to keep a list of such useful references.

To save time and effort in describing targets, entries can be made in the following form:

Land targets.   Name or position/bearing from ship/approx. height.
Ship targets.   Name or description/aspect/light or loaded, &c.
Buoy targets.   Name or position/type/topmark, 'Ra.Refl.,' &c.

This may look complicated, but is in fact very simple as the following examples show:

Land targets.   Beachy Head/NNW./350 ft.
                  Mangalore/SE/150 ft.
Ship targets.   800 g.r.t. collier/end-on/light.
                  1500 g.r.t. tramp/beam-on/light.
                  Beaverdell/inclined/loaded.
                  Owers lightvessel/end-on/—.
                  100-ft. drifter/end-on/sail set.
Buoy targets.   'Knoll'/spherical/stf. and diamd.
                  'East Cant'/can/ra.refl.

If a land target consistently gives a clear *isolated* echo on the radar screen and corresponds to a precise point on the chart, it can be termed a *radar-conspicuous object* and marked R.C.O. in the log. Conversely, an object of

Duplicate

PART I.—OPERATIONAL LOG (in Triplicate)

S.S./M.V. "*Nonesuch*" TYPE OF EQUIPMENT *Radar Ranger*

Top Copy to be retained in Ship.
Duplicate and Triplicate copies to be forwarded to the Marine Department of the Company.

SHEET NUMBER.

VOYAGE NO.	DATE	TIME ON	TIME OFF	AREA OF USE	WEATHER AND VISIBILITY	REASON FOR USE, BENEFITS OBTAINED AND LIMITATIONS OBSERVED	STATE OF EQUIPMENT	TIME SAVED
23	21.3.'49	1100	1700	Mersey & N. Ireland	Slight Sea. Vis 6 T. until 1200: Then land haze		Good: P.M. reading normal 1 hour miles, land haze obscured shore marks 1200-1700 CN	
"	22.3.'49	1000	1100	Rockall	Mod. Sea. Vis 8 miles T.		Good: range rings disappeared. Cal. unit repaired - O.K.	
"	23.3.'49	1000	1100	N. Atlantic. Long Mod Sea. Vis 8 miles T. at Noon 25° W			Good.	
"	24.3.'49	1300	1400	N. Atlantic. Long Mod Sea. Vis 12 miles T. at Noon 33° W			Perf. Mon. reading 70% normal. O.K. after changing Crystal	
"	25.3.'49	0630	0830	N. Atlantic: approaching St. John, N.B.	Calm Sea. Vis ¼ mile - fog.	W.S.L. and E.H. Enabled Ship to proceed right into Harbour entrance Scale of shortest range would have been inadequate for entry into a less directly approached harbour. Ra. Refl. on Fairway buoy most useful.	Good	} 20 hours
"	26.3.'49	1300	1750					
"	27.3.'49	0010	1130	Making landfall				

T. = Testing
W.S. = Shipping collision warning
E.H. = Entering harbour
L. = Making landfall

Total Hours Run 256

J. Kettle ....Master.

Fig. 93.

importance to visual navigation that does not show up clearly on the screen can be called a *radar-inconspicuous object*, and marked R.I.O. In making an entry of an R.I.O. it would be useful to say in what way the target is inconspicuous, e.g. 'echo isolated but weak' or 'confused with other echoes'. For entries describing weather and sea conditions the Beaufort notation will be found convenient.

Diagrams. A sketch is often a more vivid and comprehensible way of describing an observation than a long written explanation. Sheets of 'PPI blanks' can be used for recording outstanding radar pictures. The value of so recording the radar appearance of landfalls, &c., has already been mentioned. Such phenomena as interference, shadow sectors and indirect echoes, &c., all lend themselves to a pictorial explanation.

Operational log (Fig. 93)

This part of the log, as its name implies, is intended to be used for recording all occasions on which the set is operated. It should contain items of general interest which will indicate the needs of the user and any general criticism he may wish to make. The principal information which should be recorded can be classified as follows:

(1) *Periods of use.* Two important factors governing the design of equipment are the total operating hours over a long period, such as a year, and the maximum length of a single period of operation.

(2) *Areas of use,* so that those studying the log will be aware of the general conditions obtaining.

(3) *The reasons for switching on* the radar equipment will help a great deal in clarifying the needs of the user.

(4) A statement of the *benefits obtained or limitations observed,* will give the user an opportunity to point out particular merits or weaknesses which he discovers in practice. Time saved on passage by the use of radar is usually difficult to estimate, but is of much interest when it can be included.

The information inserted in this part of the radar log will be of great use to the shipowner, who will be able to assess the value obtained from the radar equipment, and to the radar designer by indicating the importance of each use of radar and how far short of requirements the present product falls.

At times the master may wish to make additional comments which do not fit into either of these sections of the log. These might be general remarks about the use of radar and its effect on the navigation of the ship, or it may be that some particular event logged elsewhere deserves a fuller report. These remarks could be put into an additional report which may or may not be considered part of the log proper.

Permanent information

The radar shadow sectors of a ship will be changed only by alterations to the ship's structure or position of the scanner. A diagram showing the shadow sectors should be exhibited near the display. The inclusion of similar information in the log, with such permanent information as type of set and modifications made, date of installation and aerial height and the maximum reading of the performance monitor will also be of great assistance to those who require such information for study.

14

SIMPLE MAINTENANCE

ALTHOUGH the modern radar equipment is designed for reliable service as well as high-class performance, it is inevitable that from time to time faults will occur and it is essential that the set should receive a certain amount of routine maintenance. On many ships an officer fully qualified in radar maintenance is available to carry out this work, but on some others, especially those engaged on short passages, maintenance is almost entirely left to the shore engineers. The purpose of this chapter is to give the observer officer first some guidance as to the regular maintenance a set should receive, and secondly some general information on the tracing of faults. This may enable him, in the absence of a qualified maintainer, to clear minor or easily rectified faults which might result in the set being out of action when particularly required. If the fault cannot be remedied, it may help him to give the regular maintainer useful information as to its nature, which will assist the latter to carry out a repair with minimum loss of time.

Details of the design of radar sets differ widely; specific information for each make of set is of course to be found in the manufacturer's maintenance handbook, but a few examples are quoted in this chapter.

Warning

It is important to remember that high voltages exist in nearly all parts of a radar set. Where there is easy access to units employing high voltages safety switches are fitted which break the appropriate circuits when the cover or door of the unit is opened. The term *high voltage* is applied to all circuits in which the direct and alternating voltages combine to give instantaneous voltages greater than 250 volts. It should be realized, however, that voltages below 250 volts may still be present in a unit even though the safety switch has operated correctly. 50 volts is regarded as the 'threshold' of danger. When working inside a unit it is advisable to switch off the main d.c. supply and, if work is being carried out at some point remote from the switch, to place a conspicuous sign on it to warn others.

Before going aloft to work on the aerial unit the officer should remove the main d.c. fuses and take them with him. A shock from a comparatively low voltage is sufficient to make a man lose his grip.

When it is intended to make a voltage measurement inside a unit, the safety switches can be made inoperative but care should be taken to ensure that meter probes have well-insulated handles.

The references to high voltages, radio frequency, radiation and X-radiation in the Board of Trade Marine Radar performance specification should be carefully studied.

SIMPLE MAINTENANCE

Routine maintenance of moving parts

The following routines should, unless otherwise stated, be carried out at regular intervals, which are usually indicated in manufacturers' handbooks.

Motor alternator. Examine the motor alternator and grease the bearings according to the manufacturer's instructions. The commutator brushes and brush gear should also be carefully inspected, and cleaned of grease or carbon dust. The carbon brushes should be free to move in the holders without being slack, and the brush spring should press firmly on and move freely with the brush.

When fitting new brushes it is important that the correct grade of carbon should be used and that, before passing current through the new brushes, they should be bedded to the commutator surface. Bedding of a complete set of new brushes can be carried out by placing around the commutator, with the rough side uppermost, a strip of carborundum cloth or fine glass paper (not emery cloth), long enough to overlap by an amount at least equal to the circumferential distance between adjacent brush boxes, so that, when the brushes are lowered on to it and the commutator rotated, the strip is tightened. After removing the carborundum cloth remove the carbon dust from the machine; this can best be done using hand bellows. Finally wipe the commutator and brush gear insulation with a clean cloth.

The commutator surface should appear polished, of uniform colour and free from irregularities. Again if necessary carborundum cloth may be used but never a coarse glass paper or emery cloth.

The static power converter requires little attention. Ventilation through the unit should not be impeded and the unit should not be subjected to excessive ambient temperature. Relay contacts should be kept clean and terminal connections secure. All components should be kept free of dust and moisture and capacitors should be periodically inspected for signs of oil leakage.

Some static converters can suffer considerable damage if the input mains polarity is reversed.

The scanner. Inspect and clean the scanner driving motor, as described for the motor alternator, and in addition check the oil level in the scanner gear box and top up as required. After every 1000 hours of use drain and refill the oil sump.

Inspect the sealed window on the feed-horn and clean off any deposit of soot. If the deposit of soot is found to be heavy, the horn window should be cleaned once a week.

The exposed hinges and screw threads on the scanner unit should also be inspected and if necessary greased.

Inspect and clean any slip rings in the scanner unit and in particular clean the heading marker contacts. Use *fine* glass paper or carborundum cloth for cleaning the contacts. Check the accuracy of the heading marker by comparing the radar bearing of an echo with the visual bearing of the target (see Chapter 3).

Trace-rotating mechanism. If the set is fitted with rotating PPI deflection coils, check the tension of the slip ring contacts and clean the ring with a dry cloth. Clean and lubricate any exposed gears on the turning mechanism.

Mechanical inspection. In addition to the inspection of moving parts it is advisable to attend to the following points during the maintenance routine.

(1) Make sure that all interlock safety switches operate correctly.

(2) Examine the units for mechanical damage to components, looseness of nuts and bolts, &c. Ensure that anti-vibration mounts are firmly secured to the units and to the ship's structure.

(3) Check that all valves and other 'plug-in' components are firmly held in their sockets; make sure that where valve-retaining devices are provided they are in good condition (tension) and in place.

(4) Check desiccators, where fitted, and replace if saturated. The silica-gel cells commonly employed in radar sets are usually impregnated with cobalt chloride, which causes them to turn pink when saturated. A saturated cell may be reactivated by baking in an oven until the crystals regain their bright blue colour.

(5) Examine any oil-filled transformers for oil leakage.

(6) Check that the range-marker and electronic bearing cursor controls operate smoothly.

(7) Check that the bearing cursor rotates freely.

(8) Check that ventilation inlets are unobstructed.

Routine electrical maintenance

Before examining the display for indications of set performance, the following electrical routine should be carried out. The proper frequency of this routine is difficult to forecast in general terms. Where manufacturers' handbooks contain precise instructions, these should be followed as far as may be practicable.

(1) Measure and record the readings obtainable from built-in meters. If any reading is outside the limits usually stated in the manufacturers' handbooks it should be regarded as a warning of impending trouble.

(2) Remove dust and other foreign matter from components and wiring, paying particular attention to high-voltage circuits. Use a small paint brush to remove the dust, and finally blow through the unit with a blower or hand bellows.

(3) Check that all inter-connecting leads are firmly secured in their terminal blocks.

(4) Examine the earth connections of screened cables to ensure that they are firmly made, particularly on flexible screened cables fitted between movable units and the frame.

(5) Examine all tell-tale lamps and operational lighting for proper functioning.

Performance checking

Performance checking is the routine examination of the display to ascertain the level of overall performance and to check that any unwanted effects are not due to set faults.

As has been stated in Chapter 2 the performance level is gauged by measuring the length of the plume of the performance monitor. Should the range reading obtained be more than 20 per cent. below the logged reference figure, it will indicate that the set is in need of attention and it should be re-tuned (see Chapter 3). There are many other possible causes of reduced performance level, some of which should have been removed in the course of the mechanical and electrical routines mentioned, while others are considered below.

The performance monitor echo box is itself liable to become defective if it depends for correct action upon an electrically operated 'paddle' or a vibrating 'reed'. If this device is not functioning the plume on the display will be broken up as shown in Fig. 94. In this case, no reliance can be placed upon the range reading obtained.

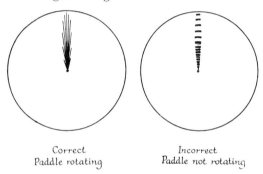

Fig. 94. *The performance monitor plume on a PPI.*

Switch to each range scale in turn and check that brilliance, focus, gain, anti-clutter, range-ring, range-marker and electronic bearing cursor controls operate satisfactorily. Check the variable range marker's accuracy on each range scale against the appropriate range rings.

Check that heading marker is present and, if azimuth stabilization is fitted, switch to compass and carefully observe heading marker. It should be possible to see the heading marker following the slight hunting of the gyro compass.

A variety of unwanted effects which may appear on the display are described in Chapter 7, which gives a brief indication of the unit in which each trouble is likely to originate. The remainder of this chapter is devoted to instructions on fault finding such as may be within the capacity of the observer and the facilities at his disposal.

Fault finding

A number of the components in a radar set, having a limited life, will fail from time to time and will require replacement. It would be misleading to specify the expected life of any components, but those which by their nature and employment are subject to ageing may be expected to have lives of the order of 1000 hours. Many have lives in excess of this figure while, on the other hand, the life of a comparatively new component is sometimes ended by the failure of another. It will be found in practice that where many valves are used their failures represent a large proportion of the total.

General advice

When removing a glass valve always hold it by the base and if necessary lever the valve gently out of its socket.

If replacing a valve in a unit does not clear the fault do not leave the new valve in position; replace the original after the fault has been remedied.

Before condemning a valve as faulty, make certain that good connections have been made, i.e. that the valve has been fully pushed home in its socket and that any clip-on connection is firm. If there is any doubt about the valve being faulty, try it, where possible, in some other unit of the equipment which is known to be functioning correctly and where the same type of valve is fitted. Replace the valve in its original position if it is sound.

A component which is suspect should be set aside and labelled with the description of the fault and the length of time it has been in use; it should be possible to extract this information from the radar maintenance log.

Without the proper knowledge and test equipment never attempt to adjust tunable coils in the receiver or calibrator units. Great care is necessary when using test equipment to check transistors, which are susceptible to damage from unduly high voltages.

A blown fuse usually indicates an overload due to a faulty component; occasionally a fuse will fail as a result of its own weakness.

When replacing a fuse, check that the new one is of the correct rating. Never attempt to rectify a fault by using a heavier fuse; this may only increase the extent of the damage. When re-wiring a fuse holder leave sufficient slack wire to allow for tightening under the screw head. Make sure the wire is well clamped under the screw. A poor connection will cause a high-resistance joint due to the wire overheating and oxidizing.

Location of faults

From the appearance or non-appearance of certain effects on the display, it is usually possible to decide which section of the equipment is most likely to be at fault. It will, therefore, be found useful to divide the set into the following sections and to memorize their exact positions.

PULSE GENERATOR OR TRIGGER UNIT.
MODULATOR AND TRANSMITTER.
R.F. SECTION AND AERIAL.
I.F. PRE-AMPLIFIER.
VIDEO AMPLIFIER.
TIME BASE.
CALIBRATOR.
VARIABLE RANGE MARKER UNIT
DISPLAY.

AFC UNIT. ELECTRONIC BEARING CURSOR.
I.F. MAIN AMPLIFIER. TRUE MOTION UNIT.
  POWER SUPPLIES FOR ALL UNITS.
  MOTOR GENERATOR.

These sections can be identified in Fig. 17, which is a block diagram intended to show the channels by which the basic units pass information to the display. It will be seen from this diagram that three principal types of information are fed to the display:

(1) That causing radial movement of the spot so as to form a trace. This involves the time base, the trigger unit, the modulator and the transmitter.

(2) That causing rotation of the trace, in synchronism with the aerial.

(3) That causing variation in the intensity of the spot to produce echoes, range rings, range marker, electronic bearing cursor and heading marker.

A fault can usually be localized into one of these channels by examining the display. The fault-finding chart overleaf shows a logical step-by-step examination, leading to a particular item in the fault list which follows it. For example, when the trace is rotating satisfactorily and range rings are present but there are neither echoes nor speckled afterglow, the chart indicates item I in the fault list, which shows that the fault lies in the i.f. amplifier. Of course, the chart and the fault list necessarily lack detail but they should enable an observer to locate a large proportion of the faults which occur in practice.

When the examination indicates a fault in a particular unit, if it contains valves, first examine them and replace any that are suspected of being defective; then examine the other components in the unit for damage and finally, if possible, check the voltages supplied to it.

The extent to which the observer will be able to make use of the guidance given in this chapter will depend very much on the individual. There is no doubt that if the principles of the radar set are understood and the approach which is suggested here is developed, the average officer will be able to contribute a great deal to the continuing efficiency of his equipment, and in addition will greatly improve his own operating ability. It is equally certain, however, that if the individual finds himself out of his depth, he will be well advised not to proceed further without expert assistance.

It may be wondered why True motion is not mentioned in the chart and Fault list which follow. In fact all the points in the list apply equally to True and relative motion sets. Any fault in the True motion facilities is likely to be in the computer/resolver unit. These differ very greatly between manufacturers and may be electronic or mechanical or a combination of the two.

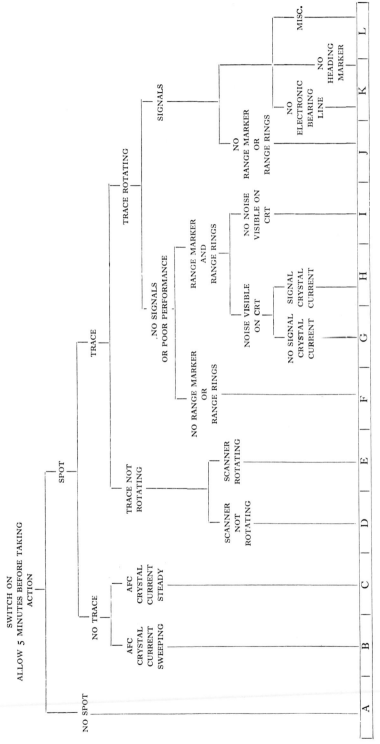

Fault-finding chart.

SIMPLE MAINTENANCE 249

FAULT LIST

(It is assumed that we are dealing with an equipment using valves rather than transistors. However, a great deal of the routine applies to one using transistors. For care in testing the latter, see Appendix IV.)

A. *No spot*

Check first that the power supply to the set is satisfactory by reading the output voltmeter on the motor generator set or control board. Check that all safety switches are closed and that all drawers and removable panels are secured in position. If the power is ON, suspect first the high-voltage rectifiers in the display unit power pack and secondly the CRT.

B. *No trace: AFC crystal current sweeping*

The fact that the crystal current is sweeping and no trace is visible on the screen indicates that the transmitter is not functioning. Check the modulator by observing the modulator valve. If it is not working suspect the modulator valve, valves in the pulse generator or trigger unit, and the rectifier valves in the modulator power pack. If the modulator is functioning correctly check that the magnetron is working by holding a neon tube in front of the magnetron output or waveguide output from the transmitter unit. If the AFC crystal current is sweeping when the transmitter is working the fault lies in the AFC unit. Check the valves in the AFC unit and the setting up procedure.

C. *No trace and AFC current steady*

The AFC crystal current being steady indicates that the transmitter is working satisfactorily and, therefore, the fault lies either in the cable carrying the sync or lock pulse to the display unit or in the time base. Check the continuity of the sync cable between the transmitter and the display unit with an ohm-meter. On most sets a small spark should be seen when the inner conductor of the sync cable is touched to earth at the display end.

Next check the fuses of the power supply to the time base generator, then the valves in that unit and finally the continuity of the deflection coils round the CRT through the slip rings.

D. *Scanner not rotating*

If the scanner is not rotating make sure the aerial switch is ON then check the supplies to the scanner motor by means of a voltmeter and examine the brushes or slip rings of the scanner motor. When a brush is removed from its holder, care should be taken to replace it in the same position with respect to the direction of rotation. In most sets the supply to the scanner motor is taken via a relay which operates only when the transmitter is working, e.g. three minutes after switching on. Check that this relay is closing and that the spring contacts are making good connections. With the power switched off rotate the motor shaft by hand to check that the scanner gearing is not seized up.

E. *Trace not rotating*

The types of bearing transmission mostly used are:

(*a*) Geared Selsyn.
(*b*) *M*-type step-by-step transmission.
(*c*) Rotating magnetic field.
(*d*) Servo system.
(*e*) Synchronous motor drive.

Failure of the trace to rotate in any of the above systems with the exception of the rotating magnetic field, may be due to a seizure of

FAULT LIST—*continued*

the gears in the PPI turning mechanism or to a misalignment of the CRT causing it to foul the deflector coil system.

If the trace fails to rotate on a set using geared selsyns, check the voltages on the stator windings and ensure that the three stators are connected together. Check also that the selsyn motor windings are the correct resistance. If the stator windings are disconnected care should be taken to reconnect them in the correct order otherwise the trace may rotate in the opposite direction to the aerial.

Should the trace fail to rotate in an equipment using *M*-type transmission, check the voltage applied to the *M*-transmitter, and connections to the *M*-receiver. Check also that the brushes on the *M*-transmitter are making good contact.

Failure of the trace to rotate when the rotating magnetic field system is used can be due to failure of the mechanical drive between the aerial and aerial magslip, or to a break in the connecting wires between the latter and the deflector coils.

In equipments using servo-systems, failure of the trace to rotate may be due to faulty valves or components in the servo amplifier, to a faulty velodyne motor, or to faults similar to those mentioned above under geared selsyns.

Where synchronous motor drive is used the results can be achieved in two ways: either

(1) The PPI and aerial turning motors are both synchronous motors fed from the same a.c. source.

(2) The synchronous motor in the PPI is supplied from an auxiliary alternator mounted in the aerial unit and geared to the aerial turning motor.

Failure of the trace to rotate will be due either to a faulty PPI motor or to connections between aerial and PPI, in the case of (1) above, or to a faulty auxiliary alternator in the aerial unit in case (2).

F. *No signals or range rings*

If neither signals nor range rings are visible, the indications are that the fault lies in that part of the equipment common to both, i.e. the video amplifier or the power supply to it. An exception to this is in sets where the signals and range rings are fed to separate electrodes of the CRT.

Check the valves in the video amplifier and its associated power supply and the connections to the base of the CRT.

CAUTION. See that the equipment is switched off before examining the base of the CRT as the voltage at this point may be between 6000 and 8000 volts. It is safest, even after switching off, to short-circuit the base connections to the chassis with a well-insulated screw driver before touching any connections.

G. *No signals or crystal current*

If the speckled afterglow is visible on the CRT but there are no signals and no signal-crystal current, suspect the signal crystal. If there is neither signal nor AFC-crystal current, suspect the klystron local oscillator.

H. *Crystal current but no signals*

If the signal crystal current and receiver noise are visible but there are no signals or performance is poor, it is probable that the receiver is satisfactory and that the fault lies in the transmitter or RF Head and/or the waveguide.

If long-range signals are good but short-range signals are missing, the most probable cause is the TR Cell as this is a typical end-of-life effect in a TR Cell (as its recovery time increases).

If all signals are missing or weak, first check the output of the

SIMPLE MAINTENANCE

transmitter by holding a neon tube over the output waveguide. Next check the tuning of the local oscillator and, where applicable, the tuning of the TR Cell. Next check the TR Cell by substitution and, if signals are still missing, the fault will probably be due to moisture in the waveguide. Disconnect and examine any portion of the waveguide where water may have accumulated!

I. *No receiver noise*

If there is no speckled afterglow (receiver noise) on the display, the fault lies in either the i.f. amplifier or its associated power supply. Check the valves, the supply voltages and the coaxial cables joining the various parts of the amplifier chain. The gain control and sea-clutter control form part of the i.f. amplifier and these controls and the connections to them should be checked. No attempt to adjust the i.f. amplifier tuned circuits should be made without the correct test equipment and a thorough knowledge of its use.

J. *No range marker or range rings*

If signals are present but no range marker or range rings, the fault probably lies in the range marker or calibrator unit. Check valves and power supplies. Do not attempt to adjust the coils in the calibrator unit.

K. *No heading marker*

If the heading marker is not showing, examine the heading-marker contacts in the aerial pedestal. Most equipments incorporate a selector switch on the display control panel which is in series with the ship's head contacts and this should also be examined in case of failure. Clean the contacts with either carborundum cloth or fine emery cloth.

No electronic bearing cursor line

If the e.b.c. line is not showing, examine the e.b.c. contacts. The moving contact is mounted in the display unit, on an 'output shaft driven by a servo motor and gearbox. The fixed contact is adjustable by an operational control. Closure of the e.b.c. contacts causes an oscillator to function and the output is passed to a squarer, which in turn gives an output of suitable shape and duration for PPI presentation. This is fed to the video amplifier. If signals and other markers are present, the fault probably lies in the e.b.c. generator, unless inter-unit wiring etc. is faulty. Where valves are used these should be checked; otherwise diagnosis requires test data and equipment for the particular type of equipment in use.

L. *Miscellaneous faults*

A short length trace is probably due to a faulty amplifier valve in the time base, or to low h.t. supply.

If a signal crystal burns out shortly after being renewed, suspect the TR cell. Before replacing the cell check that the 'keep alive' voltage is being applied.

15

THE RADAR-EFFICIENT SHIP

BETWEEN the assurance that a ship has been provided with an up-to-date type-tested radar set and the confidence that her officers will now have, always at their disposal, a highly efficient aid to safe and timely arrival there is a wide field of surmise. The factors which contribute to radar efficiency in a ship are many. Some are material and some are personal, and together they weave a story which begins in the factory and culminates, perhaps, on the bridge of a ship at sea in fog. The object of this chapter is to examine the more appropriate of them and to suggest means of ensuring that that much needed confidence will be present at all times and be firmly grounded.

The story first takes shape in the laboratory and the drawing office, where the initial and mainly irrevocable steps are taken in the production of an article which will have not only to pass the type-test of performance and durability laid down, but also to win the satisfaction of the user over long periods. This satisfaction will be based more on freedom from breakdown than it will be on superlative performance. By its very nature, radar is susceptible to breakdown through the failure of one or other of its hundreds of components due to ageing or other causes. It is probably too much to hope that valves will ever have anything but a short life in terms of years but, in modern sets, the vast majority of these have been replaced by transistors, which have a very much longer life, and one is entitled to expect that increasing knowledge of the behaviour of equipment of this kind will lead to the development of condensers, resistors, transformers and so forth, which will last for the useful life of the set.

Radar is not an end in itself but merely one of a number of means to an end. For this reason it must accommodate itself to conditions which already exist and its design must take account of the limits of seagoing maintenance skill, which are set by the likely capacity of existing personnel and the reasonable amount of training they may be spared to undertake. It is probable that, in deep sea ships at any rate, the feeling will grow that a set which cannot be kept in efficient working order by seagoing officers is not worth having. The successful set will be the seamanlike set, always prepared for emergencies. Therefore, in addition to aiming at the type test, which in Great Britain includes a rigorous series of climatic and durability tests as well as proof of high level radar performance, the designer must have long term stability and reliability as his target, which no brief test-room routine can assure. This, needless to say, implies an ease of fault location and component or circuit unit replacement within the average ship's capacity.

When the design engineer has given place to the production engineer and he to the sales manager, the story moves on to the ship and to the placing in her of this potentially valuable aid to her safety. The extent to which this potential may be realized will depend very considerably on the care with which the positions for the various parts of the equipment are chosen. If the ship is a new one, it may be that the constructors have had in mind the needs of the radar equipment, in which case the installation problem may not be formidable. But there will be many older ships in which the ideal will not be attainable.

Siting the scanner

If there are difficulties, those of siting the scanner are likely to be the most severe. It is seldom that an all-round view for the aerial is practicable, unless it is placed high aloft which is usually done in new construction. The obstructions to radar view at lower levels are funnels, masts and samson posts. In some cases it is found undesirable, for one reason or another, to place it higher than the funnel. Although the shadow will usually be thrown right aft, it may be a considerable nuisance with overtaking ships and with navigation in fog when contact with objects passing astern may quickly be lost. Samson posts present a greater problem as they are often numerous and some are likely to be relatively close to the aerial. A number of small shadow or blind sectors such as they may cause, is liable to be a considerable nuisance, and wherever possible the aerial should be placed above them. Foremasts cannot often be avoided, and the average lowermast causes an appreciable shadow. Worse than the mast itself, however, are very often the cross-trees, and if the aerial were to be placed at or just above their level a serious blind sector might be caused. If the aerial can be placed a few feet below them, this will be avoided. When other considerations make it necessary to place the aerial above the cross-trees, it should be well above, so that their shadow is brought very close in to the ship. Particular care in this respect is necessary in cargo vessels, which trim by the stern when in ballast and may thus bring the cross-trees up into the 'field of view.' This is not a simple problem and it has been found to be accentuated in many ships by the level of the cross-trees being about 8 to 10 ft. above the deck of the upper bridge, just above the level of a scanner mounted thereon.

Height, therefore, is much to be desired when siting the scanner for a good view. It is of value for another reason, because the distance of the radar horizon increases as the square root of the height. It will have the disadvantage of increasing the length and vulnerability of the waveguide and therefore the loss of power in it and the range and amplitude of sea clutter will be increased. However, with modern equipments the advantages of view and horizon will override these. The important factors of vibration and ease of maintenance will need to be taken into account.

It was mentioned in Chapter 7 that obstructions to radar view are liable to cause false (indirect) echoes; this effect can, however, often be reduced or eliminated by covering the obstruction with deflecting or absorbent material.

Whether the scanner should be sited on or off the centre line is a matter of some controversy in relation to the use of radar for anti-collision warning. It is not intended to discuss this aspect here, but it may be remarked that, before making a decision to offset from the centre line, it would be well to examine closely the effect it would have on the size and positions of shadow sectors. On the necessity for exact and intimate (by the observer) knowledge of these sectors and for exhibiting the information on a notice or tally plate near the display there is no dispute.

Protection. Although physical protection of the aerial is not necessary, it is important that its rotation should not endanger personnel. A consideration, which is sometimes overlooked with unfortunate results, is that if the movement of the aerial is arrested by loose halliards, aerials, guys, &c., it is quite possible that the scanner motor will burn out and the aerial be distorted. As the likelihood is remote of a replacement being available at any but major service depots, this is obviously a danger to guard against when siting as much as when operating.

Siting the display

The siting of the display unit or console is a matter of great importance, though it is unlikely to present any great difficulty except in the smaller and the older ships, in which the bridge compartments may be small. The immediate convenience of watchkeeping officers or of the master or pilot when they are in direct control of the ship's movements is regarded as a paramount consideration and the principal display will no doubt be sited in the wheelhouse. If the chartroom is separate from the bridge and easy comparison with the chart is important a second PPI may be placed in the chartroom. A number of subsidiary factors are involved, such as night operation with darkened wheelhouse and lighted chartroom, and day operation where screening of the display from bright daylight may be necessary. Repair work on the console at night will, of course, be easier in a lighted compartment, but this is not an overriding factor. In the wheelhouse positions, the display is often provided with curtains or screens and is usually placed so that the observer faces forward and, in some cases, so that he can maintain a visual lookout from the same position.

In many modern vessels the 'chartroom' is an integral part of the bridge, usually central and just abaft the steering position. All the navigating aids, Direction finder, echo sounder, Decca Navigator, Electric log and the radar are grouped here with adequate table space for chart work, radar plotting, &c.

Siting the transmitter

The positions of the transmitter and other units are usually chosen on the grounds of general convenience and technical preference. Thus the length of waveguide between transmitter and scanner should be as short and straight as is compatible with the requirements already mentioned. A limit is usually placed by the manufacturer on the length of waveguide and on the length of the cable run between transmitter and display. The transmitter should be sited accordingly. Ease of maintenance should not be forgotten.

The echo box, if separate from the scanner unit, is usually sited so that its plume on the PPI will fall within a shadow sector, or, if there are no shadows, on a bearing on which echoes are of least importance, e.g. right astern. It should be at a distance from the scanner which will give a plume of reasonable length (about 1000-2000 yards) and width.

Interference with other equipment

The compasses and the radio are the only items likely to be affected by magnetic or electric interference. Each portion of the radar equipment is tested for effect on compasses and is allotted safe and (sometimes) conceded distances from standard and steering compasses. Radar units should only be placed within the conceded distance in very exceptional circumstances.

It is a statutory requirement that the radar should not cause interference with the radio installation or with broadcast reception. This may affect the choice of a site for the transmitter unit, since the modulator is the most likely source of electrical interference. Needless to say, all parts of the equipment liable to cause noise or vibration (scanner, generators, blower motors, &c.) should be placed where they will not disturb officers, crew or passengers.

Operation

Having covered the main considerations of installation which will affect the ultimate potential value of the radar set, the rest of the story is in the hands of those who take it to sea. The radar shore depot organization, of course, will have a part to play, but whether it is a leading role or a 'walking-on' part will depend very much on the seagoing user. The operational use of radar may be considered to have two sides, obtaining the information and making use of it.

Obtaining the information. Anyone who has absorbed the rudiments of getting the best out of a radar set, as they are set down in the foregoing chapters, will be aware that there is much more involved than might at first appear. The complementary arts of handling the controls and interpreting the PPI cannot be fully understood without some instruction and will never be mastered without much practical experience at sea. Courses of training for observers are held at radar schools at certain ports. These courses are very valuable both for imparting the appropriate degree of knowledge of the set and for putting the radar and the navigational sides of the question in their proper perspective. If these principles are put into practice at sea, the observer can become both competent and confident.

The direct benefits of this competence are mentioned elsewhere in the book. The results of practice may be, not only an improvement in the general ability of the observer, but also the building up of a familiarity with the radar appearance of important parts of the coasts which should stand the ship in good stead when visibility is low. The value of recording results has also been stressed in Chapter 13.

Using the information. Data obtained from the PPI will only contribute to the safety of the ship if it is used with discrimination. The certainties must be distinguished from the probabilities and whatever results from this analysis must be used in conjunction with information from all other available sources. As in the art of obtaining navigational information from the PPI, clear weather practice in the use of it will pay a handsome dividend when fog blots out the forecastle head. The full implications of such effects as the distortion of coastlines due to lack of bearing resolution cannot be appreciated fully until they are seen and compared with the chart. Exercises in 'blind navigation' and 'blind pilotage' can easily be arranged with an officer standing by to ensure that no action is taken which might endanger the ship. The greater appreciation of situations which efficient plotting of ship targets will give has probably to be experienced before it will be regarded as a necessity. With practice, and only with practice, the six-minute interval for manual plotting and choice of action suggested in Chapter 9 can be reduced. Simulator courses at the radar schools have been devised to give dynamic anti-collision training to officers who will have the responsibility for handling ships in fog.

Maintenance and readiness

A quotation from the burial service—'In the midst of life we are in death'—may aptly be applied to radar. If it does break down, it is inclined to choose the most lively moments. The user will do well to ensure that his set is maintained and regularly brought to a high level of performance and to give all possible assistance to the maintainer, in the way of advance warning of operational requirements and of description of the nature of faults and of behaviour preceding the breakdown (Chapter 14). It will be found desirable to have the set warmed up and working before the time it is needed operationally. This implies that it should be at short notice whenever and wherever thick weather is likely. A clear weather, daylight river passage may become a nightmare with little warning, if the wind takes heavy smoke from factories across the channel. At least one casualty has been penalized by the Courts for failing to have radar operating before entering smoke on the Thames. It is to be hoped that a time will come when all failures will be within the capacity of the qualified shipboard maintainer to diagnose and remedy. At present a high percentage should be, but the complexity of the circuits is such that occasionally a fault defies diagnosis by seagoing staff.

In ships which depend on shore depots for maintenance the same assistance by the user will be of value. Manufacturers' service manuals usually contain instructions for simple maintenance, which can be carried out by users when no qualified maintainer is available. In one or two sets, the built-in testing facilities obviate the carrying of much additional test equipment. It is desirable that a radar-fitted vessel should carry a multi-range meter of the 'AVO' type and an instrument capable of measuring insulation resistances up to several megohms, but these will usually be for the use of the maintainer (see Appendix IV). Such instruments should only be used on transistor circuits by qualified maintainers, because of the risk of damage.

The rewards

In the radar-fitted ship, where care has been taken that each stage of this story has had a happy ending, the sequel usual in fairy tales should apply and the chances of all arrivals being safe and timely should be greatly enhanced. When time is money the radar set is often able to earn its keep. It is not always appreciated that margins of safety allowed when no radar is available can be reduced without an increase of risk when radar is used with judgment. When the tide is a governing factor at the port entrance, an hour saved by radar may mean the difference between a safe berth and an exposed anchorage. Its contribution to safety is present in almost every aspect of seafaring in low visibility. Even if, in the early stages of design and installation, the ideals have not been reached, the users, by giving attention to the details emphasized in this chapter, can turn radar to the greater advantage of their ship.

16

RADAR IN THE FUTURE

IN its peaceful application, marine radar is hardly out of its adolescence. In twenty years it may have grown out of recognition. Whether it does so or not depends upon a number of factors of which, perhaps, the most important is progressive thought on the part of the user and his employers.

The future development of radar is governed by factors which do not apply, with the same force at any rate, to most other items of ship's equipment. A shipowner is able to build ships to new designs, to improve main machinery, generators, winches, refrigeration, &c., almost as he pleases, and very often his own staff is competent to plan the new designs or specifications for him. There are also many highly competent manufacturing and consultative organizations available, with whom the shipowner may discuss such subjects should he wish to do so. Most of these items have been gradually developed over the years; the function of each is quite specific and restricted. Several of them are affected by Government and other regulations, but not to an extent which would seriously condition development. Most of them, whether efficient or unreliable, can only affect the ship in which they are installed.

Radar, it can be seen at once, is in an entirely different category. Its development has been and still is amazingly rapid. The type of radar from which the present-day marine radar sprang was unknown only six years before the major part of its development was complete in 1945. It was developed to meet the need for detecting an enemy in the air or on the sea. When radar was introduced as an aid to peacetime navigation, the full scope of its future employment could hardly be visualized. It was not designed in fact to meet a completely specified need, but was adapted to meet the obvious sea requirements which seemed to be within its scope. Use and thought over the intervening years have caused it to develop three distinct functions, navigation, pilotage and anti-collision, each of which requires of radar somewhat different characteristics, as will have been seen in previous chapters. A fourth function, berthing, has not yet been given much attention.

Although sometimes of great use in clear weather, radar comes into its own when visibility is restricted, so its development is closely linked with that of navigation in fog and the willingness of owners to encourage their masters to take the additional risks involved in blind navigation in estuary, river and port. It is in connection with this linkage that good contact between user, owner and design engineer is vital. When profit margins are minimal, companies can ill afford to have ships held up in port approaches due to being poorly equipped for movement in fog; but all the functions mentioned are highly complicated, involving arts and sciences about which the electronics engineer, initially, will be completely ignorant.

Whereas the principles of navigation may be studied in many volumes, the niceties of its practice and the habit of mind of the men whose lives are devoted to its problems may not. Collision avoidance is an even more complex subject, as it involves human behaviour under stress and techniques as yet by no means fully explored. Pilotage and ship handling in narrow waters have their own infinity of characteristics; this art, in clear weather, is a mixture of eye and memory, of anticipatory judgment and almost subconscious appreciation of movement and distance, by impressions from many sources. Radar has a long way to go before it can offer the ship handler an efficient and practical substitute for an all-round view of the land and its multitude of features, its colours and its perspective. The berthing problem in nil visibility has further complications, and will almost certainly require special equipment both afloat and ashore.

This description of the most difficult aspects of the radar navigational problem may serve to emphasize the wide difference in outlook between the user of the equipment and its designer. The closing of the gap is vital to securing the ultimate objective, which must surely be to make the mariner independent of the visibility, as well as to reduce his workload, improve his judgment and increase his spontaneity.

When there is no longer any disparity of view, there will still be the matter of cost, which, despite radar's unique potential contribution to safety, is a holding factor. Shipowners often keep radar to what they conceive, sometimes wrongly, to be the essential safe minimum and manufacturers are wary of developing revolutionary equipment for a comparatively small and hesitant market. The costs of research, design and production are high and require deep consideration, particularly when the market for conventional equipment is restricted to new ships and replacing worn out or obsolete sets. These costs are naturally reflected in the price of radar sets and the step-by-step progress, with something better available every five years or so, has kept prices, not steady, but in a kind of rut, or rather in several ruts, because there is a wide price differential between the best set and the worst. These adjectives may be used because the best may be taken to be not good enough and the worst to be severely restricted in operational efficiency and versatility.

The price differential tends to parallel the tonnage differential, which is irrational, because, when moving in fog at sea or in confined waters, a small ship may be just as great a menace to general safety as a large one or a tramp as a passenger liner. Small ships' navigation is on the whole more intricate and demands quality in radar. The best radar would be a very small part of the cost of any ship which it might be instrumental in saving from total loss.

On the other hand, the best radar will be wasted in any ship, if it requires more man-hours than can be made available. This would rule out from small ships the more sophisticated radars of today as an anti-collision aid, because they do not go far enough in taking the load of choice of targets, verification, observation and calculation from the radar observer and leaving more of his time available for appraisal. The fully automatic plotters, which would leave almost all of his radar time free for appraisal will be, at the present time, outside the price range of small-ship equipments. As the computer and the associated circuitry which form the basis of these automatic devices become more commonplace the cost may fall. The use of computers is not confined to these high-priced equipments; smaller versions, sometimes called mini-computers, have already been introduced in radar receivers such as those at Liverpool Shore radar (Chapter 12) and Raytheon TM/CA (Chapter 9 and Annex).

The advantages which arise from the use of computers stem mainly from their speed of operation which provides a very high data rate, by which is meant very frequent up-dating of the raw data (observations) and so of the computed data. Their versatility is great; they can be used to store information and to draw channel boundaries and the limits of danger areas on the PPI for safer navigation. Great developments may be expected in this field in the future.

Apart from such fundamental conditions, there are several ways in which radar may be improved. The remainder of the chapter is divided between short- and long-term development.

SHORT-TERM DEVELOPMENT

Although most commercial marine radar sets today will, at their best, give an exceedingly good performance, there are a number of aspects in which some, if not all of them, could be improved. These can be classified as follows:

The reliability of sets

The form and degree of maintenance which is given to radar sets after installation varies considerably. Some ships depend entirely on shore maintenance, while at the other extreme there are those who hope to use shore depots only for the replenishment of spare parts. Between these extremes there are many variations. Whatever form individual practice may take it must surely be a principle that, in the nature of seafaring, the ship with its available personnel and its quota of spare parts should be self-sufficient. That is to say, the reliability of sets should be such that, in ships which prefer to do their own maintenance, the staff on board can keep them in efficient operation, and in short-voyage ships which depend on shore maintenance the set should never break down at sea. While this may remain an ideal, there is no doubt that many equipments have an incidence of faults which is too high, and very often a type of fault occurs for which it would be unreasonable to carry replacements. The standard of training which is given to men who are intended to carry out shipborne maintenance is well known and is unlikely to be changed materially. So designers know the standard of maintenance which their products in those ships will receive and the kind of facilities for maintenance which the sets should offer. In parallel, therefore, with the improvement of reliability, consideration should be given to ease of fault location.

The above criticism refers mainly to equipments which use valves rather than transistors. Manufacturers have been at pains to improve the reliability of their radars and the main trend of development has been in the direction of replacing valves by solid-state devices and in the introduction of printed circuit boards and integrated circuits. This, in the last decade, has been a major step forward in radar reliability; there is still a way to go before radar can challenge, with the permanence of its availability for operational use, any other item of ship's equipment.

Several manufacturers have installed their own facilities for climatic and environmental testing of complete equipments. This is another notable advance.

Consistency of performance

Apart from the kind of defects which will put the radar out of action, there is a tendency in the average equipment today to fall a long way from its peak performance after not very many hours in use. The restoration of performance to its peak value is a matter which requires an amount of skill and knowledge unlikely to be available in personnel who have not had a course of training. Although simple means are provided for detecting a deterioration in efficiency, the user, though accepting this as a useful warning, will hardly be satisfied with the inferior performance. Here again there is a favourable trend in modern development.

Detection range and target identification

Probably every radar set of the present day is capable of obtaining echoes from land targets at the maximum range for which it is designed provided that the echoing characteristic of the target is sufficiently great. From the deep-sea navigator's point of view, however, there is a world of difference between obtaining occasional unidentifiable echoes at long range and being given a picture of part of a coastal area which he can identify with certainty from his chart, or with characteristic echoes observed and recorded on previous occasions. Identification in one of these degrees may be said to become important at distances from the nearest danger somewhat greater than the error in position which it is reasonable to anticipate. Obviously this distance will vary considerably between one occasion and another but it is thought reasonable to put it at 25 miles. This criterion can only be applied, of course, when the land is sufficiently high and responsive but it will be found that a large proportion of the better known and more important landfalls on the normal sea routes do possess the necessary characteristics. Despite this, it is believed that many installations will not reach the suggested standard at the present time. This may be faulty choice on the part of the shipowner (see below).

Requirement for good detection ranges is not confined to the long-range identification of land targets. It seems highly important that developments directed towards other objectives, such as reliability and high resolution, should not obscure the need for good detection range of smaller and closer targets. It is of great importance that objects such as small buoys should be detected at not less than 2 miles, and it is very desirable that really poor targets, such as sandbanks, should be detectable at, say, a mile.

Improvement of resolution

There is ample evidence of a trend in development in the direction of improving resolution in range and bearing and so increasing the value of radar as an aid to navigation in narrow waters. Such improvements may make the use of a larger scale display worthwhile and may affect the planning of the character and disposition of radar navigational marks which is to a considerable extent conditioned by the discriminative ability of shipborne radar. Further, the better the resolution the more nearly will echoes of natural and haphazard targets resemble their visual appearance.

Shipowners do not always take advantage of the very high degree of discrimination available, perhaps because of a disinclination to instal the largest aerial offered with the set. The bearing resolution of a 12-ft. aerial being twice as good as that of a 6-ft. aerial might well make the additional cost worth while.

Improvements in PPI presentation

Considerable advances have been made in the development of so-called 'Bright displays'. The general object is to provide a display which can be viewed comfortably in the ambient light level on the bridge without having to use screens or a visor, which inhibits viewing by more than one person simultaneously. Although probably not fully effective in noon sunlight in the tropics, they achieve the object in considerable measure. Manufacturers differ in the precise technique used, but the common object is to suppress receiver noise and radar interference so that a higher level of brightness can be employed. Special arrangements may also be made to deal with sea- and-rain-clutter. Although not yet perfect, this is a considerable operational advance.

Improvements in aids to the use of radar

Apart from the radar set itself, considerable improvement may be looked for in the provision of ancillary devices to make the use of radar more reliable and comprehensive. Although efficient radar reflectors have been produced and installed in quite large numbers, there is much to be learnt about the best method of placing them so as to facilitate identification of important seamarks. The use of these devices to improve the response of poor targets such as wooden fishing vessels and boats will no doubt increase.

One of the most valuable aids to visual navigation in narrow and difficult waters is the leading mark. Keeping two leading marks in line provides an extremely sensitive guide to the correctness or otherwise of the ship's position. The development of racons and ramarks has passed into the operational stage and perfection of this type of aid will greatly increase the value of radar in the identification of land echoes and also no doubt in the recognition of important floating targets such as lightships and pilot vessels. It may also provide a solution in some form to the problem of leading marks.

The improvement of charting technique from the point of view of radar recognition of coastal features has been given continuing attention. The expense of chart production and the labour of correction, however, make it highly desirable that the same chart should be used for both visual and radar purposes; the trend in Admiralty charting is in this direction. The development of the technique of contouring charts is of great interest in view of the help it may give to the long-range identification of targets.

LONG-TERM DEVELOPMENT

The kind of radar described in this book has already done a great deal towards lightening the burdens of those who are responsible for the safety of ships at sea. Having accomplished so much in so short a time it is perhaps natural to suggest that it should do more. It is safe to say that it can already do more than any other aid to reduce the menace and inconvenience of fog, and the inclination of the mariner will probably be to demand that it should go the rest of the way and make him entirely independent of the visibility. It may well be that radar by itself can never do this, but it seems likely that the radar principle will be involved in any complete solution to the problem.

To the mind of the navigator, conscious of the importance of a safe and timely arrival, fog brings four urgent needs:

(1) The removal of the atmosphere of 'blind man's buff' when encountering another vessel in open waters.
(2) The provision of means of fixing his position when in the vicinity of land, with a progressive speed and accuracy comparable with that which he needs and can achieve in clear weather.
(3) Pilotage facilities in waters where he would not normally fix but would work by eye.
(4) A substitute for the senses of movement and space with which, in good visibility, he confidently turns his ship 'on her heel' and takes her alongside a wharf or into a lock.

It is in connection with the first and last of these needs that the shortcomings of present-day radar are most apparent. The other two may be met in good measure by minor developments in orthodox radar.

Radar for avoiding collision

In clear weather, in circumstances where action is necessary to avoid a collision or close approach between two ships, two of the factors on which a master's confidence and judgment are based are his ability to estimate the other ship's aspect and the knowledge that he will detect an alteration of course by her almost as soon as her helm is put over. With radar, the only accurate information which is continuously and immediately available is the range and bearing. However expert he or his officers may be in the use of the manual radar plot or with the study of echo trails, he will not know with certainty, from moment to moment, the other ship's course. From his plot he can determine what this was a minute or two previously and make a plan for any action he intends to take, but it will take him at least this length of time to discern an alteration by the other ship when his own is on a steady course, and longer should it occur during or just after his own alteration. The development of automatic plotters gives the problem a different complexion. (See Annex.)

These minutes are precious, and they may be fatal. If the aspect of the other ship were apparent on the PPI a very great deal of this doubt would be resolved. It has been shown in Chapter 6 that only in exceptional and very close-range circumstances will the shape of a ship echo be governed by the shape of the target vessel, and that this is a question of resolution in range and bearing. It will be remembered that the characteristics of the radar set cause the radar beam to have a minimum area of resolution, which is governed mainly by the pulse-length and the beam-width, and that with 3-cm. radar this echo area will not be substantially enlarged and altered in shape by even a large ship target except at short range.

A pulse-length of 0·05 microseconds and a beam-width of 0°·1 would give a minimum area of resolution of about 8 yards square at the edge of the 3-mile range scale, neglecting the spot size; on a 12-inch PPI on the 3-mile range scale a $\frac{1}{4}$ mm. spot would cover a circle 10 yards in diameter and would increase the minimum area of resolution to 18 yards square. This would give a great possibility of discerning the aspect of a ship target at 3 miles. At 9 miles range on a 9-mile range scale, the area would be approximately 38 yards × 54 yards, elongated in azimuth. It might still be possible to 'see' the near waterline of a large ship target. However, the technical considerations must be examined. The best spot size obtainable at the present time is $\frac{1}{2}$ mm. A pulse-length of 0·05 microseconds is available but on the 3-cm. wavelength a 0°·1 beam-width would need an aerial 72 ft. wide! To enable a 6-ft. aerial to be used the wavelength would have to be about 2·5 mm.

Scientists would not necessarily regard these technical requirements as impossible, but it must be remembered that a radar set operating on 1·5 mm. might have a maximum range of only 2 or 3 miles and would therefore be useless for general purposes. Further, at 1·5 mm. the penetration into rain is very poor. Even with these limitations, however, such a radar might be of great use as an aid to avoiding collision in fog and its short-range performance should be phenomenal. It is possible to imagine its use as an auxiliary to the main set, employing as many as possible of the circuits of the latter and so saving some expense.

The 8 mm. radar mentioned in Chapter 1 goes a considerable distance in this direction (Plate 3). It has a beam-width of 0·15° and gives a good indication of the aspect of fairly large vessels at six miles. Its range is limited, however, and its performance in heavy precipitation would be poor.

Apart from the improvement of the quality of radar data, there is the question of facilitating its transformation into the form in which the navigator can make the best use of it. The basic principles have been fully discussed in Chapter 9. Various methods of computing were mentioned, including technical devices such as the ARP and PRP and the reflection plotter. Whether these are effective or not depends mainly upon *whether the observer has time to use them*; one must presume that he will have the inclination to do so. It seems likely that, in the vast majority of ships, those which have only one officer in the bridge watch, the chances of an effective plot being maintained up to the moment when it ceases to be of use are remote.

When there are many echoes on the screen, it is necessary to select for closer examination those which present any serious degree of threat. With any method short of complete automation this must absorb a significant amount of the observer's time. The aids to computation which are available, have been tried or are under development, fall into three categories of use:

(a) Manual echo selection followed by manual treatment;
(b) Manual echo selection followed by automatic treatment;
(c) Automatic selection and treatment of all echoes.

The whole object of the computer in this field is to divest the mariner of the need to spend time on mental and manual operations which can be done much more accurately and rapidly by automatic means. The latter will certainly be expensive and, therefore, it might just as well be made to do the whole job. It may then prove less expensive than the observer's time and the possible result of diverting his whole attention on to trigonometry at critical moments. The speed of the computer may be the crucial factor, in a collision-risk situation, in presenting the facts needed for an appraisal within the time available.

The implication is that the computer should display its output graphically on the PPI in the form of true or relative vectors, at choice and immediately it is required. These must be displayed for all echoes simultaneously and the current positions of echoes must not shift when changing from true to relative or vice versa. This would entail a permanently 'centred' display, which could always be heading upwards.

This collision-avoidance equipment is mentioned in this chapter because the newer, more sophisticated systems have not by any means found general acceptance yet. It may be that further development will take place before the conservative seafarers and ship-owners find sufficient confidence to allow electronics to stand in for the overstressed manual operator and give him time for appraisal before it is too late. (See Annex.)

In connection with this use of radar it may be wondered why no mention is made of the possibility of providing devices for indicating to other vessels that radar is being used and for transmitting some form of identification signal as a prelude either to course and speed data or to establishing communication. There is a number of reasons.

The efficacy of the Rule of the Road as a means of preventing collision is founded on three virtual certainties. They are that those in charge of ships at sea have good eyesight, an ability to use a compass and sound experience of the Rules. To add to these any assumptions based on the efficiency of performance and competence in employment of a series of electronic devices would be a matter requiring the gravest consideration.

Safety at sea is an international matter and the circumstances of ships meeting at sea in fog usually contain an element of urgency which will not brook the delays inseparable from identification, establishing radio communication, coding and translation.

When more than two ships are in proximity, Rule of the Road problems are apt to be complicated, even in clear weather. When depending on radar for information they will probably be more so. The situation likely to develop if all ships seek to use their communication systems in order to hear what others are saying and doing or to talk their ways out of difficulty is best left to the imagination.

Radar as an aid to berthing

There are quite a number of vessels which use normal marine radar for berthing purposes, and sets with similar characteristics but installed on shore are used to maintain ferry services in fog. The short-range performance characteristics of most modern shipborne radars is good enough to be used for this purpose in many cases. These, however, must be regarded as special cases, in which familiarity with the locality and assiduous practice in the interpretation of the radar picture are the over-riding factors which inspire confidence. The degree of confidence built up in the mind of the master or pilot is the ultimate factor which results in the ship anchoring in the approach or proceeding to berth.

There is plenty of evidence to show that the onset of fog in the approaches to ports which have to be entered by narrow and winding channels is the cause of much congestion and immobilization of shipping. No doubt the difficulties are accentuated by the presence of ships which have no radar working and must therefore anchor or trust entirely to the guidance of the port radar station, but it can be foreseen that, when the efficiency of both shipborne and shore-based radar has been increased to an extent which will make blind pilotage commonplace, the port authorities will see to it that the channels are kept clear for those ships which are prepared to complete or commence their voyages in fog.

There will, of course, be occasions on which a sudden decrease in visibility will cause ships to anchor in channels, but this, it must be hoped, will be the exception. On the other hand, this possibility combined with lack of confidence in radar is one of the present deterrents to the continued progress of large vessels, for which anchoring in the channel or even at the inner end of it may be hazardous. This thought must be taken a little further, to the point at which it appears more hazardous to anchor than to proceed with the aid of radar and so the assurance that, on occasions, the development of better radar will be in the interest of safety as well as of time-saving.

In addition to the exceptional cases mentioned earlier, radar is frequently used all over the world to leave port in fog when the channel is straightforward and when the process of unberthing and shaping for the channel can be carried out without its aid. Similarly, the operation of entering a port is often accomplished with radar, if the inner anchorage is open or visibility within is sufficient for visual berthing. These manœuvres will, of course, be greatly facilitated by improvements in the short-range performance of radar either on the orthodox equipment or perhaps by the use of millimetre waves as mentioned. The closer a ship is brought to her berth, however, the greater will be the urge to complete the voyage and it is really in the final berthing operation that the greatest confidence is needed and that present-day radar, even with short-term improvements, will fail to establish it.

An entirely different approach to the subject of berthing (not only in poor visibility) has been inspired by the advent of the VLCC and other giant vessels and the need to berth them alongside without damage to themselves or to the jetty. In one case this takes the form of two doppler radars mounted at the berth opposite the bow and stern of the ship. They will measure movements as small as 4 ft. per minute. The operating frequency is 14·1 GHz. The aerial is a parabolic dish which can be trained and elevated.

The visual case

In the introduction to this chapter a brief impression was given of the complicated processes going on in the mind of the ship handler when taking his ship in clear weather to a berth in narrow waters. The expert in such matters is sometimes credited with a sixth sense, which may perhaps be construed less mysteriously as Anticipation. Intelligent anticipation in this sense is merely a forward estimate based on appreciation of rate of change. The ship handler is concerned almost exclusively with the rate of change of his position, which is conveyed to his mind as a complex assortment of rates of change of bearings and ranges measured by his eyes on a wide variety of fixed targets. He is concerned in fact to possess an instantaneous and continuous appreciation of movement over the ground, which he obtains very largely by watching the transits of fixed objects, usually in the horizontal plane though occasionally in the vertical. His selection of transits depends upon the direction in which he is moving or wants to move or, particularly, does not want to move; he may be watching three or four of these at the same time and may change them minute by minute. The eye is minutely conscious of perspective and of changes in it. At times this may be the operative factor without any conscious selection of targets. The pilot is interested also, of course, in actual bearings and ranges, but more particularly in the way they are altering or not altering, and in the width of navigable water available round the ship. His interest in the ship's heading is usually incidental and is concerned more with its relation to the land he is approaching and to the tidal stream rather than to any particular point of the compass.

The confident ship-handler identifies himself completely with his ship. It is as if he himself were moving, turning and being influenced by set and drift. To place him in a position of remotely controlling his ship through the medium of a small-scale diagram is to break the cord of synthesis. The difference is fundamental, and calls to mind the analogy of back-seat driving.

In fog

What substitute for this sensitive performance by the eyes can radar offer the master or pilot? The outstanding shortcoming of the radar display today is its complete inability to convey any instantaneous impression of movement, let alone the rate of movement. The view offered on a 9-inch PPI is comparable with that from an aircraft at 20,000 ft. The rates involved in a berthing may be as little as a tenth-of-a-knot (two inches per second) and these are easily discernible to the eye of the ship's pilot by the visual methods described. To be able to identify a particular object in the usual clutter of radar echoes from the land is most unusual so that the use of haphazard transits is virtually impossible. A painted advertisement passing behind a chimney pot may be a most useful visual transit. Radar is at some disadvantage here, with its tiny two-dimensional picture in monochrome.

It is not intended to suggest that there is only one way of obtaining the necessary information or that radar will fail if it cannot conform. No doubt the ship-handler can learn other methods, but before he will establish confidence in them he must be convinced that his moment-to-moment control of the ship is not impaired.

If familiarity with the circumstances can breed sufficient confidence in some people for them to berth their ships with the aid of present-day equipment, it should not be impossible to find means of building it up for those who do not have the advantage of constant practice.

If, as has been suggested, the all-important senses are those of appreciation of movement and space, the problem for radar would seem to be essentially, though not exclusively, one of scale. Since the important aspect of movement is not that through the water but that relative to fixed objects, improvement in scale will need to be accompanied by improvement in the delineation of shore features, natural and artificial. The need for an appreciation of space is not so much a desire for exact distances as for a relation between the size of the ship and the amount of safe water around her. A considerable increase in scale would imply that the ship herself would occupy a not inconsiderable part of the PPI picture and could be shown artificially upon it, thus offering a direct comparison.

Any substantial increase in scale, which might have the effect of making slow movement discernible in a single brief observation, would however at once introduce a major disadvantage if the radar display technique remained as it is today. Unless True motion was used and kept in perfect adjustment, the tube afterglow would smudge and destroy the clear outlines of the front of any target whose range happened to be increasing and of one side of any target whose bearing was changing. To avoid this in shipborne radar it would be necessary to adopt a much shorter afterglow and a much more rapid presentation of the picture, i.e. a much higher speed of aerial (and trace) rotation. Whatever method of scanning was adopted, better discrimination would be needed. The degree of improvement suggested in the previous section on Collision Warning might suffice.

On the picture presented by a shore-based radar, the only echoes which have afterglow trails are, of course, those of moving targets. This suggests that, if the ship were able to see the shore radar's PPI, the smudging difficulty at least would be removed. Further, a shore-based radar can be free of some of the limitations imposed on shipborne radar, such as aerial size and the need for long-range performance; thus the shore radar can be developed for very high resolution, and could no doubt be placed at various points within the port area from which the most significant views would be obtained, e.g. the new Teesport system will have four radars at three well separated sites. With these advantages the shore radar picture would display many refinements unattainable by the shipborne set. Such a picture displayed on the bridge of a ship in port approaches and/or wishing to berth in fog should go far towards building up the confidence needed. Means have already been found of transmitting such a picture to the ship for display on a suitable PPI or possibly on that used for normal working.

Another system, on which operational trials have been carried out, transmits a television picture of the shore radar display. To receive this, only a television receiver is necessary in ships. It introduces various problems, of course, including that of self-identification in thick weather when position is uncertain.

As mentioned in Chapter 12, Television display technique is being developed and is already used in a number of Shore Radar Stations. The purpose is to convert the radial and rotating scan of the PPI into the horizontal lines of the TV raster. Among other advantages, this provides a better contrasted display for daylight viewing. The conversion may be done by using a TV camera to photograph a small (4-in. perhaps), high-definition PPI and displaying the result on a TV-type display (as at Teesport. 1978). In some French stations a scan conversion tube is used which does the conversion within a single envelope.

This is a developing area of the subject and one which may extend to shipborne equipment.

In this connection it is believed that the aim of any such equipment must be to give the man on the ship's bridge *information rather than instructions*, since he alone is in a position to make the critical decisions based on a knowledge of all the circumstances.

When examining these possibilities, consideration will no doubt be given to the extent to which the ship should be independent of shore assistance. The greater the dependence and the more complicated and costly the equipment the fewer will be the ports at which it is available.

It is probably true to say that any device which is merely an aid to navigation will, if it is to prove generally acceptable for use in merchant vessels, need to fulfil these requirements among others:

> Its cost must be commensurate with its value either in reducing unavoidable risk or in saving expense due to delays.
> Its reliability must be unquestionable and must take realistic account of training possibilities.
> It must not require super-specialists in operation or maintenance, or additional personnel.
> It must give its information rapidly and intelligibly and, as far as possible, in a form to which the navigator is accustomed.

The main objects of this chapter have been to suggest directions in which the long-term development of marine radar might proceed, and to emphasize the importance of the user in encouraging these advances. Responsibility for safe navigation is a heavy and a very personal one. In any description of how it is discharged, the word judgment would figure largely. This being so, it is inevitable that there will be, for some time at any rate, widely different opinions on the requirements and their relative priority. This difference will exist, not only between one type of trade and another, but also between identical ships on the same route. It is important, therefore, that information and opinion should flow in from the sea, so that as many requirements as possible may be satisfied and that in time the sailor may regard fog, not with apprehension or irritation, but as an interesting exercise in the conjunction of two sciences—navigation and electronics.

Annex

COLLISION AVOIDANCE SYSTEMS

THE book remains a valid statement of the principles and of the simpler aspects of marine radar technology and also of its use at sea. Since the first edition was published in 1952, considerable changes in equipment have occurred and many of them are mentioned in the later editions. However, what is perhaps the most fundamental change in the last seven years has been the introduction of the computer as a plotting aid. When the last revision of the book was made in 1968, this, in seagoing form, was still below the horizon.

To give some perspective to developments in commercial marine radar, the first thirty years of its life may be divided into three equal periods. The first decade was devoted mainly to producing and improving radar to meet ships' needs; reliability, PPI diameter (increased from $6\frac{1}{2}$ inches to 15 inches), operational facilities, compass stabilization, bearing resolution (aerial beam-width), anti-clutter devices, reflection plotter and the introduction of true motion.

In the second decade this general development was continued but, perhaps because collisions were still occurring between ships fitted with radar, more attention was given to reducing the load on the operator of observation and computation in the hope of accelerating the delivery of intelligence in usable form to the Master. The result took a variety of forms: ship's-head-upward stabilized display, semi-automatic transfer of echo position to plot, photographic projection of the display on to a plotting surface, etc., but none of these proved to be universally popular.

In the third decade, which is only just ending, events began to move faster. The term Collision Avoidance Radar came into common use and several radars were produced in which plotting, or the claimed substitute for it, formed an integral part. The Decca 66AC, the Marconi Predictor and the Raytheon TM/CPA were in this category and appeared in the early part of this period; the Kelvin Hughes Situation Display appeared somewhat later but qualifies for inclusion.

Some of these equipments and some of the considerations involved are mentioned in Chapters 9 and 16. More individual attention to the principles will be given later in this annex. Towards the end of the second decade and at the beginning of the third, engineers and others experienced in the computer field and in its application to military and civil projects such as missiles, air traffic control, lunar probes, etc. became interested in improving on the, to them, archaic methods employed at sea in the vital task of converting radar data into directly usable form. The companies primarily concerned were IBM, Iotron, Norcontrol, Raytheon and Sperry, all basically US firms except for the Norwegian Norcontrol. Other companies have joined this list and Raytheon has now temporarily left it.

We may therefore distinguish three successive contributions to the present state of the art:

(i) *Normal plotting methods* (reflection plotter, etc.).
(ii) *Aids to appraisal* (electronic markers but no computer).
(iii) *Computerized systems* (automatic tracking and processing of data).

The plotting controversy

Largely owing, no doubt, to the considerable cost differential, some controversy has developed over the respective merits of the computerized plotter and more conventional systems. Most published criticism of the former has come from the manufacturers of the latter and vice versa; the shipowners, with notable exceptions, have been hesitant, possibly on account of the wide variety of advice proffered to them by their normal radar manufacturers, the makers of the computerized automatic systems and their own technical and professional advisers. When facing radar costs of a different order from those to which they have been accustomed, though not an escalation very different from that in the field of building and operating ships, they would naturally hope to hear a consensus of informed opinion on one side or the other; this is not easy to obtain on a subject which covers such a diversified interface of sciences and professions, while the concept of cost effectiveness is difficult to apply to a subject in which the hoped for end-product is negative i.e. avoidance; and so much depends upon human capacity, performance and veracity, all of which are individual imponderables.

To consider the controversy dispassionately, a study is needed of the radar plot and its desired end-product based solely upon the viewpoint of safety of ships, and this study should include devices or systems which claim to provide a substitute for plotting. No authority has yet defined this end-product in specific terms, although the courts have, on occasions, castigated ships for 'failing to make proper use of radar' and even 'failure to plot'. In at least one case the court called expert witnesses to testify regarding the time-scale of manual plotting. In 1977 the United States Administration, apprehensive of continued collisions and pollution of the environment in its own waters, drafted for public comment regulations which would make mandatory in those waters the carrying and use in the larger ships of 'Collision Avoidance Systems'; here 'use' includes 'continuous evaluation' of the echoes of ships which present a collision risk. The draft regulations include a performance specification for collision avoidance systems. Many, however, would prefer to treat such systems as a valuable evolution rather than a mandatory requirement at this time.

The radar plot

It has already been made clear in Chapter 9 that the basic purpose of a radar plot is to discover which of the ships whose echoes are on the radar screen present a threat, potential or actual, to the safety of own ship. Knowledge of a threat will need to be supplemented by information regarding its degree and urgency and other data which will assist in deciding on a course of action. It seems to be agreed generally that the most effective indication of a threat lies in the predicted distance of the CPA and the time to reach it, while the most important factor in the choice of manoeuvre, if any, is the other ship's true course and speed. This basic purpose has, of course, the corollary that all significant threats need to be watched to see whether they change or remain constant, which leads to the expression 'continuously monitored'. As will be seen later, this requirement becomes of the greatest significance when ships are approaching the CPA and time is running out, which is also the period in which sudden changes of course and speed are often made, by one or both ships.

The importance of the time factor. When Chapter 9 was written, radar plotting was in its infancy and nothing more sophisticated than the manual plot had been thought of. Not many collisions between ships using radar had been studied and the importance of the time factor had not been fully realized. It is evident now that the time taken to plot one or more echoes and up-date the plot and hence to 'monitor continuously' is perhaps the most significant of all the factors involved although varying interpretations may be given to the term 'monitor continuously'.

Time available. This consideration gives the basis for starting an examination of the effectiveness of the various plotting methods and of substitutes for them. The first step is to compare the total time available for what may be called *Escape* with the time absorbed in observing the data, computing by whatever means are available and deciding on the action needed, if any; which may be called *Planning Time*. This comparison may be set out as a simple equation: *Time Remaining* for a disengaging manoeuvre equals *Escape Time* minus *Planning Time*.

A study of twelve collisions chosen at random, which have been considered in Courts of Law or Official Enquiries, shows that the time which elapsed between realization of imminent danger of collision and the collision itself, which we have called the 'escape time', was 5 minutes or less in six cases and 7 to 10 minutes in the other six. In most of the cases the initial detection was earlier, but apparently gave no cause for concern. In nine of the cases, one or both ships were altering course during the approach, which would in most cases have required a fresh start to the plot on each occasion. It may be worth recalling the categories of causes of human error (not necessarily in radar operation) tabulated in the study by the National Research Council of the U.S.A. entitled *Human Error in Merchant Marine Safety* (1976):

Panic or shock
Sickness
Drunkenness or drug influence
Confusion
Inattention
Incompetence
Anxiety

Fatigue or drowsiness
Negative transfer of training
Negligence
Ignorance
Calculated risk
Fear

Time taken in plotting manually. It is generally accepted that a dependable manual plot requires three points in line, that is, three separate observations of the echo, and that successive observations should be separated by about 3 minutes. Allowing a period for computation, delivery and appraisal, the time required for a complete plot of a single ship will be 8 or 9 minutes. This we have called 'planning time'. The length of the interval between the observations is necessary because of the slowness of the movement of the echo on the PPI; with a 12-inch PPI on a 12-mile range-scale, the surface scale is 1:144,000, so the echo of a ship doing 12 knots will take $2\frac{1}{2}$ minutes to move $\frac{1}{4}$ inch ($\frac{1}{2}$ mile). It will be seen that a three-point plot takes a lot of time and this deficiency applies largely to any system which depends on human observation of the echo movement. Up-dating the information will take the same length of time if any change in the movement of either ship has taken place and about half as long if it has not. These times take no account of the possibility of having to interleave the observations with those of another ship.

The quantitative read-out. This examination would be a simple affair if one had only to compare the performances of manual plots with that of the automatic systems. However, that would be to assume that the only acceptable end-product is a numerical read-out and, as has been said, no official statement has been made which would justify such an assumption. There are implications in its favour but, on the other hand, type approval has been given to radar equipments which include what might be described as 'aids to appraisal' but not a full plotting process, for which the British radar performance specification does not cater.

The value of a quantitative read-out may be a matter of opinion and where several items of information have to be absorbed simultaneously a pictorial presentation has its advantages, but provided the system permits the read-out to be up-dated at very short intervals it will enable changes in a target's true or relative movement to be assessed rapidly, and accurate assessments to be made as between one target and another, which may be vital to own ship in crowded waters. If the up-dating interval is short enough the result will approximate to 'continuous monitoring' and, what is more important, the information may continue to flow in during alterations in the movements of either or both ships. In the language of the computer-based systems it provides a 'digital read-out': a finite statement relating to a particular target which provides a fixed point, however short-lived, in a relatively moving world. It is something to hold on to, from which to draw conclusions.

Graphical display. A graphical display, perhaps by vectors (see below) will give the same information although not with the same precision: it may have the advantage of displaying all echoes simultaneously, but detection of a change of movement may be less immediate than with a numerical statement. However, it will give a cogent picture of the development of the situation as a whole and provide clear evidence of an increase or decrease of collision risks with individual targets.

Vectors. A vector is a graphic means of describing movement, either relative or true. In this particular connection it is named by the time interval it represents, e.g. a 6-minute vector. It is an electronic line on the PPI whose starting point is the current position of the target echo, its direction is the direction of movement of the target, either true or relative, and its length is the distance run on the scale in use in the time selected. The length is, therefore, a measure of the target's speed and the end of the vector is the position which the target will reach by the end of the time interval. It should be emphasized that all these representations are *predicted* and presume that no alterations of course and/or speed will be made in the time interval selected. In some systems the time interval is fixed, in others its length is under the control of the operator. When true vectors are displayed, own ship's movement is shown by a vector in exactly the same way.

Discontinuities in the plot or in the supply of data for it. An important cause of delay in the delivery of computed information, by whatever means, is discontinuity in the supply of data to the plot; even in automatic systems it is the running average of from five to ten observations that is updated at every scan; other examples are breaks in continuity occasioned by such instrumental functions as resetting true-motion display to avoid loss of view ahead, resetting Image Retaining Panel, re-starting recording history on magnetic tape. Then there are the long intervals between observations when plotting manually (effectively discontinuities) and re-starting the plot whenever either own ship or target alters course or speed.

Each of these causes a hiatus while the echo moves sufficiently for any new change of movement to be recognized and measured with reasonable accuracy. The length of this break in the 'readability' of the information depends upon many things but in manual plotting is unlikely to be less than three minutes.

Any system which depends for all or part of the assessment of risk in a maritime encounter upon visual discrimination on the PPI is bound to suffer some delay for this reason.

PLOTTING SYSTEMS

Enough has now been said about plotting to make it possible to assess the capabilities of the various systems. It has to be emphasized, however, that not enough is known about the time factor in potentially dangerous encounters or indeed about collision avoidance philosophy in general to justify an unqualified verdict. The broad principles of radar plotting as described in Chapter 9 still provide a valid basis for assessment and this, coupled with the wider considerations mentioned above, should help to explain the aims of the designers of the equipments.

The practice employed in this book has been to avoid specific mention of particular equipments or manufacturers. However, the few modern equipments to which reference must be made are so different from one another that it would be impossible to preserve anonymity.

Manual plotting

The manual plot, whether or not assisted by devices such as RAS Plotter, Reflection Plotter, Bial Plotter, etc. does not need further reference here. There is one exception, however, the Raytheon TM/CA, which can be classified as an aid to manual plotting and merits description.

Raytheon TM/CA. A new series of 'Mariners Pathfinder' radars has been designed by the Raytheon Company of U.S.A. They describe it as a bright 'daylight viewing' display. It has a signal processor which suppresses receiver noise and enhances the brightness of weak echoes to one level and strong echoes to an even higher one. The control of sea-clutter and rain-clutter is improved since the clutter is shown at the lower brightness level while targets in the clutter will appear as bright spots when the gain is increased and STC or FTC employed. A signal correlator may be added which rejects radar interference. The suppression of interference and noise makes the background much blacker which, with the enhanced brightness of the wanted echoes, improves the contrast and gives a 'bright display'.

There is also a true-motion unit and a micro-processor, mini-computer, which operates eight markers, individually. The markers appear as bright spots which can be placed each on a different echo. Each marker has two components, one which continues to mark the point at which the plot originated and the other which moves from the origin in sympathy with own ship's movement (called OSD—own ship's displacement-marker). Thus, the two components and the current position of the echo provide the three corners of the velocity triangle. An electronic digital clock is provided which indicates the plot time from the origin for each echo separately when selected by the operator. To facilitate measurement, the point of origin of the electronic bearing cursor (e.b.c.) can be moved to any point on the PPI. Placing it on one of the markers enables the true or relative course of the target to be read off. The e.b.c. also carries a strobe which operates in step with the variable range marker (v.r.m.) and permits distances to be read off at the same time.

The display is fitted with a reflection plotter on which manual elaboration of the automatic skeleton plot may be made.

It will be seen that, although the OSD marker is automatically progressed, it is necessary to wait for the echo to separate sufficiently from the markers to permit a reasonably accurate visual manipulation of the e.b.c. to be made. As far as time consumed is concerned, the plot is still manual in character with the same delays and discontinuities—alterations of course or speed, resetting true motion, etc. There is no delivery of quantitative information without the intervention of the operator.

Aids to appraisal

This title has been used to connote equipments which are unconventional in their displays or methods but would not be classed as aids to plotting. Two will be mentioned, Decca Clearscan with AC display and Kelvin Hughes Situation Display.

Decca Clearscan with AC display. This provides a bright true-motion display through two video processors, V.P.1 and V.P.2, the former improving the treatment of sea- and rain-clutter and the latter dealing with the removal of receiver 'noise' and radar interference, bringing all echoes up to a high uniform degree of brightness and 'stretching' (in range) all echoes outside $2\frac{1}{2}$ miles on the 12-mile scale and above. This last feature makes small echoes more readily visible. The AC display provides a novel operational feature. This consists of five separately controlled line markers, one inch long, and having a bright spot at one end. Each marker's bright spot can be positioned on a target; the direction of the line is straight towards own ship. The markers remain in the same positions relative to own ship until altered by the controls; they are produced by the interscan method and so are continuously visible. A closing echo on which a marker has been placed will move either straight down the marker or will diverge from it. In the former case, the target's bearing will be constant, so it will be on a collision course. If the echo diverges, it may not be obvious immediately whether or not the CPA will be at a safe distance. By eye or by using the parallel line bearing cursor, the operator should be able to estimate the range at CPA by imagining a line drawn from the bright spot of the marker, through the current position of the echo and continued on to the point of closest approach to own ship. The accuracy of this assessment depends, obviously, upon the target having been on a steady course and speed since it was marked; to be sure of this, it will be necessary to give fairly continuous scrutiny.

When own ship alters course or speed, all target markers will need to be re-started on the echoes; when a marked target alters course or speed, that particular marker will need to be re-started. After re-starting all or a particular marker, there will be a hiatus in the appraisal until the echo has moved a sufficient distance to permit a reasonably accurate assessment of the new course and/or speed and this may take a minute or two.

If no quantitative statement is required about target course and speed or CPA distance and time, the operator's visual assessment from the true-motion echo trails and the process described above may be regarded as sufficiently reliable; if not, recourse will need to be taken to manual plotting, the time scale of which has already been referred to.

Kelvin Hughes Situation Display. This provides a bright, true- or relative-motion display with a very short afterglow. It is a stabilized ship's-head-upward display which will show either true or relative echo trails. The method of generating the trails is unusual: the radar picture is produced on a 3-inch non-persistent CRT and is projected on to an image-retaining panel (IRP) which has a persistence of 3-4 minutes. The IRP is scanned by a television camera and the picture thus obtained is projected on to the bridge display. When true trails are required, the IRP is moved at a rate proportional to own ship's speed; its direction of movement corresponds with own ship's head. For either true or relative trails the IRP is rotated with alterations of own ship's course. Obviously the forward movement of the IRP is limited and when it has reached about half the maximum travel possible it is reset automatically to the start. The operator is given warning of the approach of the reset moment and, if this is likely to be an awkward one, the operator may override the automatic reset for a further period. Obviously this can only be done once. The reset interval is 3 or 6 minutes, depending upon the range-scale in use and the override gives the same for the extension.

The Situation Display is therefore, in essence, an enhanced true- or relative-motion display. The position of own ship is fixed at the centre of the scale and the true or relative mode is shown by the direction of the echo trails. Bearing and range measurements are by electronic bearing cursor and range rings. Simple plotting on the tube face is possible and an erase button is provided to restart the build-up of trails on the IRP when intending to start a plot.

The build-up of trails has to re-start after any break in the continuity caused by change of mode or of range-scale, after using 'erase', and, of course, whenever the IRP resets. Although the trails will begin to show quite soon, it will take about three minutes for them to reach proportions which will permit new courses and speeds to be assessed. Quantitative information can only be obtained by measurement, calculation and manual plotting. The time scale, whether plotting or not, will be very much that of the manual plot.

AUTOMATIC AND COMPUTERIZED SYSTEMS

A considerable number of manufacturers are interested in this field of development, although only a few are in a quantity production, namely, IBM, Iotron, Marconi, Norcontrol and Sperry. The Marconi Predictor system qualifies as an automatic plotter of the history-recording variety but does not provide the versatility of the fully computerized systems in its method and end product, and requires a separate description. There is, however, a sufficient degree of similarity between the computerized systems, MABS (IBM), Databridge (Norcontrol) and Digiplot (Iotron), to make a detailed description of each unnecessary.

Auto tracking of targets by a computer is common to these three systems and to Sperry CAS, while echo acquisition may be either manual or automatic and there are various types of vector display. Digital read-out is provided for various items (speeds, CPA, etc.) and in some systems there is also an Auto Alarm based on CPA. Display may be the conventional synthetic picture on its own or superimposed on a live radar (intrinsic). Some systems provide for trial manoeuvres.

Marconi Predictor

This system displays, automatically, a continually up-dated 3-position track following the current position of the echo; it does this for all echoes on the screen simultaneously. The total duration of the track is $1\frac{1}{2}$, 3 or 6 minutes and so the intervals between the four points are $\frac{1}{2}$, 1 and 2 minutes respectively. The presentation is permanently centred and will show either true or relative tracks. There is no display of quantitative information and any that is needed has to be measured by the operator e.g. target course and speed, CPA distance and time and trial manoeuvre prediction. The time taken for a change of track mode is 10 seconds; the tracks are up-dated every 10 seconds and, if using, say, the 6-minute track length, the points on it are 2, 4 and 6 minutes old. If either the track length or the range-scale is altered, recording on the magnetic tape will have to be renewed; the length of the hiatus will depend upon the track length selected.

The most recent models of Predictor have improved brightness on the display due to the incorporation of noise and interference suppression and improved clutter control. The time taken to detect a change of course by a target is likely to be a minute or so; to measure it for certain will take at least two-thirds of the track length. To detect and measure a change of speed will take much longer.

MABS, Databridge and Digiplot

The following description applies wholly to Digiplot (Plate 52), with which system the writer is most familiar, but is also to a large extent applicable to the other two.

Echo selection. This is fully automatic in Digiplot with supplementary manual selection and rejection; in some systems it is manual and some provide both. Where automatic selection is the normal method, the object is to make the whole process from observation to production of computed information completely independent of the operator; known colloquially as 'Hands Off' operation. If echo selection is manual, the operator has the essential and continuous duty of being sure that his choice of targets is correct.

Echo surveillance. The video output from the radar receiver is taken to a video processor which discriminates between 'ship-sized' echoes and larger ones which will be from land and passes both to the computer. The video processor also rejects receiver noise and deals with clutter.

Echo observation and computation. The computer passes the land echoes to the display, examines the ship-sized echoes and in Digiplot ranks the nearest 200 in order of threat and selects the closest 40 for tracking and computation; the selection is checked every scan. The data is updated every scan (3 secs) and read by the computer as required. For the bearing, the computer measures from the echo pulses the angular width of the ship-sized target, the bearing being the azimuth of the final pulse less half the echo width. The range is measured at the near edge of the final azimuth pulse. The internal accuracy of the range measurement is claimed to be within 250 ft. and of the bearing is equal to that of the gyro (10 minutes of arc for a stepper output).

Echo quality and quantity. Precautions are taken to ensure that only good solid echoes reach the display and that it does not become saturated by a superfluity of echoes. To be acquired for automatic plotting a target must return at least one echo every six scans. If it does so and is within the closest 40, tracking will commence; the process takes about 30 seconds. Tracking will stop if the criterion ceases to be met but will recommence if the echo re-establishes itself. The operator is able to reject any single echo which is not within the threat alarm area.

On occasions there may be a profusion of echoes somewhere in radar range which could upset the computer's selection of the most important 40. To prevent this happening, limit lines are provided; these are chordal lines which can be placed on the PPI at any range and at any angle. They inhibit any echoes from beyond them from reaching the computer. They do not operate in the 'Look Ahead' sector, 11 deg. either side of right ahead. To prevent acquisition of false echoes from sea-clutter, two controls are provided: a Minimum Tracking Range and a Minimum Acquisition Range. Naturally, the settings of both require to be kept to the effective minimum. No echoes are shown beyond the near edge of land, except in the 'Look Ahead' sector, which permits the plotting of ships on the far side of bridges in harbour approaches.

Display of observed and computed information. There are two forms of display; graphical and numerical. Through a unit called the Analog Display Driver (ADD) the computer provides a complete, synthetic picture of the land echoes and up to 40 ship-sized echoes, which can be displayed by itself on a bright, non-persistent PPI or superimposed upon a live radar PPI display in the relative mode known as 'intrinsic'. The controls provided are the same in each case; the mode, vector length, limit lines, range scale, and the various demands for information are under the operator's control.

On the synthetic display the ship-sized echoes in their current positions are shown as small circles and their direction and rate of movement by vectors as described above, either true or relative as may be desired. A land echo is shown as a series of dots, one every two degrees, along the near edge. The character of the controls is similar to that of the radar, their functions are engraved on them and the actual setting is always visible. The trace origin (own ship) is permanently centred, the mode in use being clear from the vectors.

Trial manoeuvre. Two important operational controls are the trial manoeuvre facility and the future position control. The former permits a proposed new course and/or speed to be inserted, whereupon, on pressing the trial manoeuvre facility switch, the relative vectors of all targets will move to the directions appropriate to the change. The time taken would be that of the actual manoeuvre speeded up thirty times; own ship's dynamics are built into the operation and, if desired, a projected delay in executing the manoeuvre may be catered for.

The future position control. This permits the vector length to be manipulated. Its length depends upon the time interval selected by the operator with the control, which has a minute by minute adjustment. The length is thus not only a measure of the speed of the target, its outer end represents the target's position at the end of the time interval. Taking the radar picture as a whole, the vector ends show where each target will be after that lapse of time. By manipulating the vector length, an operator can therefore examine closely how a situation is likely to develop and how close target ships are likely to approach own ship or one another.

Accuracy. It is important to be aware of the standard of accuracy which the display can offer. As with the manual system, all deductions from the plot depend upon accurate and repeated readings of target range and bearing, accurate observations of time and accurate performance of the gyro and the speed log, the outputs of which are used in the computations and for which agreed standards are now urgently required. The accuracy of the computer itself will not in practice be a limiting factor. The effect of errors in the data, particularly errors in bearings, will reduce the accuracy of the information displayed but frequent updating and data filtering enable these effects to be averaged out.

Digital read out. The numerical display (quantitative) is known as the 'digital read-out'. It gives observed and computed information about individual (operator selected) targets. The items are paired as follows: Target bearing and distance, Target true course and speed, Target CPA distance and time and own ship's course and speed. Each pair is selected separately as required, is available instantly and can be left in view as a monitor. The Digiplot specification claims an accuracy of to within one degree in bearing and of 0·1 or 0·2 miles in range. Target true course is said to be indicated within from 1 to 3 degrees and target speed to within 0·1 to 0·5 knot.

The time factor. From first acquiring an echo at, say, 17 miles (the maximum acquisition range) it takes about 2 minutes to build up a complete vector. However, the range-scale set on the PPI is not likely to be longer than 12 miles and, therefore, when a ship echo comes within this range it will already have been acquired and tracked and will be vectored as soon as it is displayed. As has already been said, the raw data is up-dated every scan (3 seconds); this permits a change of target course to become delineated within 15 to 30 seconds of its completion; about five scans will be enough to show the alteration commencing. It also permits the plot to continue during alterations of course and speed by either own ship or target(s); this implies additionally that the movements of an individual target can be quantitatively, closely and continuously monitored when required. After an alteration by either ship, the updated information will be available within a few seconds of steadying. The total time absorbed up to the moment of offering the complete and specific information to the Master should not exceed that taken in pressing the switch(es) demanding the information needed, say 5 seconds, plus about 15 seconds after the manoeuvring target has steadied on its new course. The initial information may be appreciably affected by errors in the source data, such as gyro errors. This should be apparent to the operator as a result of initial fluctuations in the vectors or digital read out.

Collision risk alarms. The automatic warning system is an important feature of the automatic plotter. In one or two collisions it has been apparent that the bridge staff was not alert to the proximity of collision danger, possibly due to preoccupation with other tasks, such as radar navigation. Audio and visual warnings are provided which operate automatically when the predicted CPA distance of a target falls within the limit set by the operator. They may well increase the time available for evasive manoeuvre since they operate from the time an echo is acquired for plotting even if the target is outside the maximum range in use on the PPI. The audio warning may be switched off by the operator but will operate if any other target infringes the set limit. The visual alarm may be a flashing light or extra brightness of the vector, or both. A time to CPA may also be set to avoid the alarm being triggered by targets at too great distances.

Sperry CAS System

In its philosophy and end-product this system differs significantly from the others. Whereas Digiplot, MABS and Databridge indicate in the conventional manner where and how fast each of the potentially interesting ships, including own ship, is going, the Sperry system indicates where own ship should not go. On a bright, true-motion course-upward display a number of targets, manually acquired up to a total of twenty, are shown in their present positions, and also a predicted point of collision (PPC) at which the target and own ship would meet if own ship were to alter course directly towards that position. In addition to calculating this prediction, the computer calculates an area round it, based upon an operator-selected CPA distance and allowing for the size of own ship and for possible instrumental errors. This forms an elliptical area (circular for a stationary target) which is drawn on the PPI by the computer and is called a predicted area of danger (PAD). A line joins the target to the PPC, representing target true course, and a dashed portion of this line represents the next 6 minutes of its travel. Should it not be possible for the ships to collide whatever course own ship steers, (the target remaining on her present course) no PAD will be drawn but the 6-minute vector will be a solid line. The PADs show where own ship should not venture and where she may steer if course has to be altered for any reason. The essential difference between this arrangement and the vector display of the other three automatic plotters is that in them the vectors all terminate at the same point of time, and so indicate the predicted situation as a whole at that time, while the PADs represent positions at various unrelated times. Obviously, the Sperry 6-minute vectors all terminate at the same moment; their length in time is not variable. Sperry provides a digital read-out as an option; this gives the same target information for a single selected target as do the other systems. The heading line and bearing cursor line are divided into 6-minute segments to facilitate measuring speeds, etc. The synthetic picture is superimposed upon a live radar picture.

After manual acquisition of an echo a preliminary, hatched 6-minute vector will be displayed in about 24 seconds. The time taken to establish a PAD is about 90 seconds. If a second target, lying in the same azimuth as a previously tracked target is acquired, the time lags given above will be doubled; for a third target on the same bearing they will be trebled. The maximum number of targets on the same bearing which can be acquired is four. However, there is the option to add a second electronic tracker which will increase the tracking capability to 40 targets. In this case the additional time lags mentioned will be reduced by half.

ADVANTAGES AND LIMITATIONS OF AUTOMATIC EQUIPMENT

The potential advantages arising from these new developments, both for timely decision making and in relieving the work-load of the navigator, are evident and may be summarized from the manufacturers' specifications for their equipment as follows:

(i) Automatic acquisition of echoes and check of echo selection.
(ii) All relevant echoes (in some systems the nearest 200) will be examined simultaneously and the data updated at every scan.
(iii) The nearer echoes (up to 40 in number with some systems) will be displayed and vectored simultaneously.
(iv) Demands for predicted positions of echoes or other information such as range, bearing, course, speed or CPA of targets will be met instantly.
(v) Trial manoeuvres are speeded up 30 times and incorporate own ship's dynamic characteristics.
(vi) There will be no discontinuities arising from re-setting processes or alterations in course or speed of own ship or of target.

(vii) Elimination of human error in the mechanical task of plotting.
(viii) Collision Risk Alarm based on CPA distance selected by operator and independent of range-scale in use.

As with all new concepts, the introduction of these automatic systems must be followed by a period of strict and impartial testing, and, where necessary, technical modification in the light of experience, to establish general confidence in their efficiency and reliability. There can be no doubt that sufficiently advanced electronic techniques are now available, based on recent defence equipment developments.

Possible shortcomings suggested by some preliminary evaluations attempted in the U.K., the U.S.A. and the Netherlands, include:

(i) Loss of some targets fully observable with raw radar.
(ii) Permutation of tracks and confusion of vectors in dense traffic.
(iii) Discrepancies between visual observation in clear weather and the automated radar plot.
(iv) Unreliable information on near targets due to weather and clutter.
(v) A rather low mean time between equipment failures.

One factor to be taken into account in evaluating such systems will also be the capacity of the navigator to make proper use of sophisticated equipment and of the wealth of data at his command. This will be largely a matter of specialized training and practical experience. Over-reliance on a system could lead to a false sense of security and hazardous encounters.

CONCLUSIONS

The developments in shipborne radar since the first edition of this book appeared in 1952 reveal significant progress in the processing and plotting of the raw data, leading to a more timely presentation of the situation with which the mariner is confronted. These developments may be summarized under three headings:

Manual plotting methods, which typically require 3 minutes for two observations and 6 minutes for three observations plus perhaps a further 2 minutes for making calculations, drawing vector triangles, etc, are clearly of limited application (e.g. single-target encounters in open waters).

Appraisal, using electronic markers but not a computer, gives some information much more quickly than manual methods but not so quickly, completely or conveniently as the automatic methods. For example, the Decca AC system gives its information in about 1 to $1\frac{1}{2}$ minutes according to the vessel's speeds but the markers have to be reset for each new appraisal of the situation.

Automatic plotting methods, giving continuous target course and speed information which is never more than 30 seconds out-of-date, can clearly be used in a much wider range of situations, e.g. when the time to CPA is short or one of the targets in an encounter makes an unexpected course-change at a very late stage. Automatic alarm facilities are also used.

The more recently introduced computerized systems for collision avoidance promise not only a lighter work load for the navigator in times of stress and a more timely warning of impending danger, but a fuller and more up-to-date and objective presentation of the data on which he must make his decisions and a facility for assessing the outcome of any intended manoeuvre.

These systems are still in process of operational evaluation at sea and while mandatory specifications, or requirements for their installation and use in certain categories of ships or in certain waters, may be premature, there is an urgent need for agreed standards of accuracy in relation to the data inputs provided by the radar, the speed log and the gyro-compass.

If new and sophisticated equipment is not to invite over-confidence it is essential that those who use it should be fully aware of its function and limitations and highly trained in its proper use as an aid to safety in navigation.

Appendix I

ECHO RECOGNITION TABLE

The following table is intended as a guide to recognizing echoes and other effects on the radar screen. Both the appearance and the detection range of objects on the screen are likely to vary between fairly wide limits, so that the information is intended merely as a useful approximation of what may be expected.

Description	Likely appearance on screen	Checks on recognition	Probable detection ranges for 50-ft. scanner in standard atmospheric conditions
NATURAL LAND FEATURES			
Low-lying sandy coastlines	Smooth narrow line indicating water's edge	Comparison with chart. Short detection range	1–5 n.m.
Cliffs	Line varying in regularity and thickness depending on contour and steepness	Comparison with chart. Good detection range if high and square on to the ship	For 200-ft. cliffs can be up to 20 n.m., but variable
Hills and mountains	Patches of echoes may indicate general contour; hinterland may be shielded	Comparison with chart. Echoes may correspond with peaks or escarpments	15–40 n.m. depending on character. Very variable, but see diagram (Fig. 49)
Groups of islands and rocks	May give single or separate echoes depending on distance apart and resolution of set; those in background may be shielded	Comparison with chart	
ICE			
Sheet ice	Area free from echoes except at edge	If sea clutter present and adjacent, ice can be checked by contrast	
Growlers	Small, weak, irregular echoes; may be difficult to perceive in clutter		Up to 3 n.m.
Frozen pack ice	An area of small, strong, closely spaced echoes	Identity in position of paints from scan to scan	
Bergs	Individual echoes	Probable wide variation in echo strength with aspect. Slow movement	3–15 n.m.

Description	Likely appearance on screen	Checks on recognition	Probable detection ranges for 50-ft. scanner in standard atmospheric conditions
SEA			
Tide-rips	Generally a curved line; may resemble a coastline	Likelihood of occurrence, e.g. near river mouth. Most noticeable in moderate sea	½–1 n.m.
Overfalls	As patches of clutter	Reference to chart	Up to 2 n.m.
Shoals, sandbanks awash and reefs awash	Similar to clutter except in calm	Reference to chart	Up to 4 n.m.
ARTIFICIAL LAND TARGETS			
Piers, breakwaters	Strong echoes, as plan outline	Comparison with chart. Usually very prominent	5–10 n.m.
Buildings and built-up areas ..	Patches of echoes, often very strong	Rectilinear patterns or other regularities in features	Dependent on height above sea level; good targets
Roads, railway lines	Continuous bright or dark lines	Comparison with chart or map	1–3 n.m.
Isolated lighthouses	Individual echoes	Reference to chart	5–10 n.m.
Docks	As plan outline, but much subject to shielding	Reference to chart; unmistakable at short range	Up to 5 n.m.

	Nearer edge of bridge as plan outline; further edge indistinct	Reference to chart. Targets on far side may be shielded	Up to 5 n.m.
Bridges across rivers ..			
VESSELS			
Small wooden boats ..	Small individual echoes	Distinguish from buoys by movement or by reference to chart	1–4 n.m.
Fishing vessels ..	Small individual echoes	Presence of groups of echoes	Drifters, 3–5 n.m.; trawlers 6–9 n.m.
Light-vessels ..	Small individual echoes	Reference to chart and lack of movement	⎫
Typical 1000-ton ship	Small individual echoes; aspect may show at very short range	Movement, if under way; detection range	⎬ 6–10 n.m.
Typical 10,000-ton ship	Small individual echoes; aspect may show at short range	Movement, if under way; detection range	10–16 n.m.
Typical 50,000-ton ship	Individual echo; aspect may show at medium range	Movement, if under way; detection range	16–20 n.m.
BUOYS			
Very small buoys ..	Small individual echoes	Reference to chart; lack of movement; variability except in calm	½–1 n.m. (dependent on shape)
Medium (2nd class) buoys ..	Small individual echoes	Reference to chart; lack of movement; variability except in calm	2–4½ n.m. (dependent on shape)
Large buoys ..	Small individual echoes	Reference to chart; lack of movement; variability except in calm	4–6 n.m. (dependent on shape)
Reflector buoys	Small individual echoes	Reference to chart; lack of movement; variability except in calm. Echo strength in calm sea may be reduced at certain ranges	6–8 n.m.

Description	Likely appearance on screen	Checks on recognition	Probable detection ranges for 50-ft. scanner in standard atmospheric conditions
Sea Clutter			
Calm sea	Negligible echoes		
Choppy sea	Multiplicity of small echoes round ship	Extent of clutter will be greater to windward	½–1 n.m.
Very rough sea	Multiplicity of small echoes round ship	Extent of clutter will be greater to windward	2–4 n.m.
Weather Echoes			
Rain squalls	Soft edged patches but nearer edge may be sharply defined	Movement; variations of shape	3–10 n.m.
Intense rainstorms	At long range, small single echo or small patch of echoes	Movement; variations of shape	10–20 n.m.
Thunderclouds	Similar to rain squall	Movement; variations of shape	2–5 n.m.
Snowstorms	If heavy, patches as for rain squalls	Movement; variations of shape	
External Effects			
False or indirect echoes	If large, may be obviously distorted in shape compared with true echoes	Maintenance of same range but different bearing compared with true targets; may occur particularly in shadow sectors	Less than that of true targets
Multiple echoes	Probably of similar shape, but smaller, than original echo	Will appear on far side of target, at twice, three times, &c., range	Up to 1 n.m. (target range)

Side-lobes	Arc of echoes on either side of a true echo. May form a continuous ring	Turning down gain will eliminate	Up to 3 n.m.
Interference from other radars ..	Numerous dots or radial dashes, often in a pattern such as spirals	Will be most conspicuous on long range scales; will not repeat position on successive scanner revolutions	Will appear to full extent of long range scale
Second-trace echoes	If from land, may often form a V-shaped line of echoes, with point of V towards centre of screen	Appearance in positions known not to be occupied by land targets; if of ships, courses will be distorted. Will be shifted radially if p.r.f. is altered	May originate at very long ranges

Appendix II

MARINE RADAR PERFORMANCE SPECIFICATION, 1968

A Performance Specification for a General Purpose Shipborne Navigational Radar Set

1. OBJECT OF THE EQUIPMENT

To provide an indication in relation to the ship of the position of other surface craft and obstructions and of buoys, shorelines and navigational marks in a manner which will assist in avoiding collisions and in navigation.

2. RANGE PERFORMANCE

(a) Under normal propagation conditions, when the radar aerial is mounted at a height of 50 ft. above sea level, the equipment shall give a clear indication of:

   (i) Coastlines
   At 20 nautical miles when the ground rises to 200 ft.
   At 7 nautical miles when the ground rises to 20 ft.

   (ii) Surface objects
   At 7 nautical miles on a ship of 5000 g.r.t.
   At 3 nautical miles on a fishing vessel of length 30 ft.
   At 2 nautical miles on a General Purpose Conical Buoy. (The General Purpose Conical Buoy used in type testing is a 10 ft. diameter heavy duty general purpose buoy fitted with a cage and lights having an echoing area of approximately 10 sq. metres.)

(b) The equipment will be considered as capable of fulfilling the above conditions if under normal propagation conditions:

   (i) For equipment operating in the 9300–9500 megahertz band the echo of a G.P.C. buoy, whose range from the radar scanner is 2 nautical miles, is visible on the radar for 50 per cent of the scans when the aerial is mounted at a height of 50 ft. above sea level.

   (ii) For equipment operating in the 2900–3100 megahertz band the performance is at least 5db above that specified in (i).

3. MINIMUM RANGE

When the radar aerial is mounted at a height of 50 ft. above sea level the echo of a G.P.C. buoy shall be visible on the radar for 50 per cent of the scans down to a range of 50 yards on the most open scale provided, when it shall be clearly separated from the transmission mark. Without subsequent adjustment of the controls, other than the range selector control, the echo of the buoy shall remain visible at a range of 1 mile.

APPENDIX II

4. RANGE ACCURACY

The means provided for range measurement shall be fixed electronic range rings or fixed electronic range rings and variable range marker. Fixed range rings shall enable the range of an object, whose echo lies on a range ring, to be measured with an error not exceeding $1\frac{1}{2}$ per cent of the maximum range of the scale in use, or 75 yards, whichever is the greater. The accuracy of range measurement using a variable range marker, if provided, shall be such that the error shall not exceed $2\frac{1}{2}$ per cent of the maximum range of the displayed scale in use or 125 yards, whichever is the greater. Direct reading of the range shall be provided. No additional means of measuring range shall be provided whose accuracy is less than that required of the variable range marker.

5. RANGE DISCRIMINATION

On the most open scale appropriate, the echoes of two G.P.C. buoys shall be displayed separate and distinct when the buoys are on the same bearing 50 yards apart and at a range of 1 mile.

6. BEARING ACCURACY

Means shall be provided for obtaining quickly the bearing of any object whose echo appears on the display. The accuracy afforded by this means of measurement shall be such that the angle subtended by any two objects whose echoes are separate and distinct and appear on the edge of the display may be measured with a maximum error of $1°$.

7. BEARING DISCRIMINATION

To achieve adequate bearing discrimination and minimize the occurrence of spurious echoes due to radiation outside the main beam, the aerial radiation pattern in the horizontal plane shall conform to the following specification as a minimum requirement, the figures relating to one way propagation only:

	Position relative to maximum of Main Beam (deg.)	Power relative to maximum of Main Beam (db)
Main Beam	±1	−3
Side Lobes { within	±2.5	−20
	±10	−23
outside	±10	−30

8. ROLL

The performance of the set shall be such that targets remain visible within the range limits laid down in paragraphs 2 and 3 when the ship is rolling $\pm 10°$.

9. RADIO FREQUENCY AND POLARIZATION

The radiation which shall consist of either horizontally or vertically polarized waves must be substantially confined within the bands internationally agreed for marine radar below 9500 MHz and must not cause harmful interference to services outside these bands. A facility for introducing circular polarization may be added if desired; it must be possible to switch it to linear from the display position.

NOTE: At present, racons in the United Kingdom only operate on 9300–9500 megahertz horizontally polarized waves.

10. SCAN

The scan shall be continuous and automatic through 360° of azimuth at a rate of not less than 20 r.p.m. in relative wind speeds of up to 100 knots. The aerial must start and run satisfactorily in this wind.

11. DISPLAY

The equipment shall provide a satisfactory main plan display of not less than $7\frac{1}{2}$ inches effective diameter, and it must be possible, if necessary by simple switching, to operate this as a satisfactory relative motion plan display. The display shall be capable of being viewed in indirect light by two persons simultaneously without undue restriction of the angle of view. Optical magnification may be employed provided the display so magnified remains within the accuracy limitations of other Clauses of the Specification. When optical magnification is employed to attain the minimum diameter of display permitted in this Clause, the magnification device shall be a fixed component of the display.

12. SCALE OF DISPLAY

The equipment shall provide at least six scales of display whose ranges of view shall be either (a) $\frac{1}{2}$ or $\frac{3}{4}$, $1\frac{1}{2}$, 3, 6, 12 and 24 or more nautical miles, or (b) $1\frac{1}{2}$, 3, 6, 12, 24 and more than 24 nautical miles. Range scales in addition to the six specified may be provided but where provided must comply with the clauses specifying range and bearing accuracy. On a $\frac{1}{2}$ or $\frac{3}{4}$ nautical mile range scale range rings shall be displayed at intervals of $\frac{1}{4}$ nautical mile. Six fixed range rings shall be provided on each of the other scales of display; but where a variable range marker is provided on any scale of display the fixed range rings available on that scale may be at intervals of $\frac{1}{4}$ or 1 or 4 nautical miles as appropriate.

Where the facility of a continuously variable range of view is provided, the maximum range of view so obtainable shall be 6 nautical miles. On this variable range scale fixed range rings at intervals of $\frac{1}{2}$ nautical mile and a variable range marker shall be provided.

Positive indication of the range of view displayed shall be given except that, where a continuously variable range scale is provided, it need only be given at the mandatory ranges of view covered by that range scale. Positive indication of the interval between range rings shall be given.

The natural scale of display shall not be less than 1:933,888.

13. HEADING INDICATOR

Means shall be provided whereby the heading of the ship is indicated on the display with a maximum error not greater than $\frac{1}{2}°$. The thickness of the displayed heading line shall not be greater than $\frac{1}{2}°$. Provision shall be made for switching off the heading line and shall take the form of a spring-loaded switch or equivalent device which cannot be left in the 'heading marker off' position.

14. AZIMUTH STABILIZATION

Means shall be provided to enable the display to be stabilized to give a constant North-upwards presentation when controlled from a transmitting compass. This facility may be either an integral part of the equipment or an additional component which shall be capable of being readily fitted to the equipment. When a transmitting compass is used to control the orientation of the picture or of any part or parts of the display unit, the accuracy of alignment with the compass transmission shall be within $\frac{1}{2}°$ with a compass rotation rate of 2 r.p.m.

The equipment shall operate satisfactorily for relative bearings when the compass control is inoperative or not fitted.

15. SEA OR GROUND STABILIZATION

Where this facility is incorporated in an equipment the following additional conditions must be met:

(i) Transference of Speed:
By timing the movement of the trace origin over a distance of not less than 50 per cent of the radius of the scale in use or half an hour, whichever gives less time, the error in speed when compared with the speed calculated from the input signals or the setting of the manual speed control shall not exceed 5 per cent or $\frac{1}{4}$ knot, whichever is the greater. This accuracy shall apply in using any of the range scales provided for True motion.

(ii) Transference of Course:
By measuring the movement of the trace origin over a distance of not less than 50 per cent of the radius of the scale in use or half an hour, whichever gives less time, the error in course when compared with the compass input or the setting of the manual course control shall not exceed $3°$.

(iii) Circular Re-setting
The motion of the trace origin must stop when 25–40 per cent of the radius of the display remains between it and the edge of the displayed picture. Automatic resetting is acceptable.

(iv) Re-set Warning
When re-setting is by manual control, a visual warning shall be provided on the display unit. An audible warning may also be provided which can be switched off when not required.

16. OPERATION

(i) The set shall be capable of being switched on and operated from the main display position.

(ii) The set should be suitable in all respects for operation by the officer of the watch. The number of operational controls shall be kept to a minimum and all of them shall be accessible and easy to identify and use.

(iii) The design shall be such that it is impossible for an operator to cause damage to the equipment by misuse of the controls.

(iv) The time required for the set to become fully operational after switching on from cold shall be no longer than four minutes. A standby position shall be provided so that the set may be brought to the fully operational condition from the standby position within one minute.

17. POWER SUPPLY

(i) The set shall be capable of normal working under the following variations of ship's power supplies:

A.C. Variation from nominal voltage ± 10 per cent
Variation from nominal frequency of ± 6 per cent

D.C. Variation from nominal voltage:
110/220 v. D.C. $+10$ per cent, -20 per cent
24/32 v. D.C. $+25$ per cent, -10 per cent

(ii) Provision shall be made by means of fuses, overload relays, or other means, for protecting the equipment from excessive current or voltage.

18. ELECTRICAL AND MAGNETIC INTERFERENCE

(i) Electrical Interference

All reasonable and practicable steps shall be taken to ensure that the equipment shall cause a minimum of interference to radio equipments such as are customarily installed in merchant vessels. Spurious radiation in the frequency band 50 kHz to 300 MHz should be such that the maximum excursion of the voltage waveform at the output of a test receiver (as measured on a cathode ray oscilloscope) does not exceed twice the maximum excursions caused by the continuous receiver noise of the test receiver at ambient temperature.

(ii) Magnetic Interference

Each unit of the radar equipment shall be marked with the minimum distances at which it should be mounted from a standard and a steering magnetic compass. These distances shall be understood as being measured from the centre of the magnet system in the compass bowl to the nearest point of the unit in question, and are defined as the distances at which each unit, whether working or not, will not impair the functional use of the compasses. These distances shall be determined by the Admiralty Compass Observatory in accordance with the standards specified in C.D. pamphlet No. 11D. The term 'unit' is defined as any part of the equipment supplied or provided by the manufacturer.

19. PERFORMANCE CHECK

Means shall be provided for the operator to determine readily a drop in performance of 10 db or more, while the set is being used operationally, relative to the calibration level established at the time of installation. The necessary calibration information shall be immediately available to the operator in the form of a calibration label fixed on or near to the display unit.

20. CLIMATIC AND DURABILITY TESTING

The equipment shall comply with the provisions of 'A Performance Specification for the Climatic and Durability Testing of Marine Radar Equipment' issued by the Board of Trade as Appendix B to the present Marine Radar Performance Specification. Those parts of the equipment that are intended to be installed in a working space in the ship, e.g. radio office, chart room, enclosed bridge or wheelhouse, shall be regarded as 'Class B' equipment. Those parts of the equipment that are to be installed in exposed positions shall be regarded as 'Class X' equipment. Each unit of equipment shall be marked 'Class B' or 'Class X' as appropriate.

21. MECHANICAL CONSTRUCTION

In all respects the mechanical construction and finish of the equipment shall conform to good standards of engineering practice, and the design shall be such that all parts of the equipment are readily accessible. Mechanical noise from all units shall be at a level acceptable to the testing authority. Units which will be mounted in or near operational areas of the ship must not add appreciably to the noise level in these areas.

22. PROTECTIVE ARRANGEMENTS

(a) High Voltage Circuits

The radar set shall incorporate all such isolating switches, door switches, means for discharging capacitors and/or other approved devices as are necessary to ensure that access to any high voltage is not possible. Alternatively, the equipment shall be so designed that access to high voltages may be gained only by means of a tool such as a spanner or screwdriver; and a warning label shall be prominently displayed within the equipment.

The term 'high voltage' shall be taken as applying to all circuits in which the direct and alternating voltages (other than radio frequency voltages) combine to give instantaneous voltages greater than 250 volts.

(b) Radio Frequency Radiation

The power level at which radio frequency radiation becomes a hazard to personnel is specified in 'Safety Precautions relating to Intense Radio Frequency Radiation', issued by the G.P.O., and published by H.M.S.O. The radiation level from the radar must be measured and if necessary a minimum safe distance from the aerial must be stated in the handbook, which shall include instructions regarding precautions against radio frequency radiation hazard.

(c) X-radiation

External X-radiation from the equipment in its normal working condition must not exceed the limits laid down in British Standard Specification 415.

When X-radiation can be generated inside the equipment above the level of B.S.S. 415, a prominent warning notice must be fixed inside the equipment drawing attention to an appropriate paragraph in the handbook.

23. MANUFACTURERS' LIMITATIONS

Where limitations are known to exist, the manufacturer shall propose for consideration by the testing authority the maximum and minimum distances by which units of the equipment must be separated in order to comply with the requirements of this Specification. The manufacturers shall take such measures as are necessary to ensure that any such limitations as the testing authority may approve as reasonable, and any other limitations which that authority may impose, are observed in installing the equipment on merchant ships.

24. MAINTENANCE AND OPERATING MANUALS

The manufacturer shall cause to be provided with each equipment, such information as is necessary to enable competent members of a ship's staff to operate and to maintain the equipment efficiently. Maintenance manuals shall, where the type testing authority deems it necessary, incorporate in a prominent position the instructions regarding special precautions against hazard from radio-frequency and 'X' radiation (see 22b and c above).

25. ANTI-CLUTTER DEVICES

Satisfactory means shall be provided to minimize the display of unwanted responses from both precipitation and the sea.

Appendix III

SOME CONSTANTS, FORMULAE AND USEFUL DATA

THE following constants have been used in the book:

Nautical Mile .. .. 6080 ft. (1853·2 metres).
Velocity of propagation of radio waves.. $2·998 \times 10^8$ metres per second (161,800 n.m. per second, or 328 yards per microsecond).

Note.—The usual approximation for the velocity of radio propagation is 3×10^8 metres per second. A conversion factor of 3·2808 has been used for metres to feet.

Horizon distances

The dip of the horizon is defined as the angle between the horizontal plane through the observer's eye and the direction of the horizon. Where h is the height of eye in feet or H the height of eye in metres, ignoring refraction, the value of true dip in minutes of arc is expressed by the formula

$$d = 1·06\sqrt{h} \text{ (or } d = 1·93\sqrt{H})$$

which also gives the distance of the sea horizon in nautical miles.

Apparent horizon. In standard atmospheric conditions terrestrial refraction increases the horizon distance by about 8 per cent. The formula for the distance of the horizon thus becomes

$$d = 1·15\sqrt{h} \text{ (or } d = 2·07\sqrt{H})$$

Radar horizon. Radio waves, like light waves, are also refracted by the atmosphere and for 3-cm. waves in standard atmospheric conditions an increase of about 15 per cent. over the true horizon is assumed. The distance of the radio horizon is expressed in the formula

$$d = 1·22\sqrt{h} \text{ (or } d = 2·20\sqrt{H})$$

where h is the height of the aerial in feet.

Wavelength and frequency

The relation between wavelength λ, frequency f, and velocity of propagation c, of electromagnetic waves is expressed in the formula $c = f\lambda$, where c is in metres per second, f is in Hertz and λ is in metres.

Voltage, current, resistance and power

The current I which will flow in a conductor of pure resistance R when an electromotive force or voltage E is applied across its ends may be found from the formula

$$I = \frac{E}{R}$$

When the electromotive force is measured in volts and the resistance in ohms, the current will be in amperes. Strictly speaking this formula applies only to direct current working.

Whenever current flows in a d.c. circuit, power is dissipated, usefully and otherwise. The total amount of power W taken from the source of supply is given by the product of the voltage of the source E and the current I flowing into the circuit. The relationship is expressed by $W = EI$, when E is in volts, I in amperes, and W in watts.

Power ratio (decibels)

The decibel scale is a mathematical device which is used to express conveniently the ratio between widely different powers and voltages. It is useful in dealing with such problems as the amplification of echo power or the loss of power in wave guides.

The unit employed is a decibel (db) and the ratio between one power P_1 and another P_2 is expressed in db's as:

$$10 \log_{10}\left(\frac{P_1}{P_2}\right)$$

Thus if P_1 is one million times P_2 their ratio is expressed as

$$10 \log_{10}(10^6) = 10 \times 6 = 60 \text{ db}$$

and P_1 is said to be '60 db up' on P_2. If P_2 is one million times P_1 the ratio is expressed as

$$10 \log_{10}(10^{-6}) = -60 \text{ db}$$

and P_1 is said to be '60 db down' on P_2. Since power is proportional to (voltage)², the formula for voltage ratios is

$$2 \times 10 \log_{10}\left(\frac{V_1}{V_2}\right)$$

The following table shows the equivalents in db's of various power and voltage ratios:

Power ratio	db	Voltage ratio
1	0	1
2	3	$\sqrt{2}$
4	6	2
10	10	$\sqrt{10}$
100	20	10
1000	30	$\sqrt{1000}$
10,000	40	100
100,000	50	$\sqrt{100,000}$
1,000,000	60	1000

Metric equivalents

1 nautical mile	1853·18 metres
1 fathom	1·829 metres
1 yard	0·9144 metres
1 foot	0·3048 metres
1 inch	25·4 millimetres
1 metre	3·2808 ft.
1000 metres (1 kilometre)	3280·8 ft.
1 square inch	6·451 sq. cm.
1 square foot	0·0929 sq. m.

APPENDIX III

The fusing current of various sizes of copper and lead wire:

Fusing current (amps.)	Copper		Lead	
	Diameter (inches)	S.W.G. (approx.)	Diameter (inches)	S.W.G. (approx.)
1	0·0021	47	0·0081	35
2	0·0034	43	0·0128	30
3	0·0044	41	0·0168	27
4	0·0053	39	0·0203	25
5	0·0062	38	0·0236	23
10	0·0098	33	0·0375	20
15	0·0129	30	0·0491	18
20	0·0156	28	0·0595	17
25	0·0181	26	0·0690	15
30	0·0205	25	0·0779	14
40	0·0248	23	0·0944	13
50	0·0288	22	0·1095	12
60	0·0325	21	0·1237	10

Appendix IV

USEFUL TEST EQUIPMENT

WITH the aid of fairly simple instruments and the manufacturer's handbook it is often possible for the observer to localize a fault to the valve or some other component. In many cases comprehensive facilities for measuring voltages and observing pulse shapes are built into the equipment. In others, comprehensive and valuable test points are made accessible but no built-in means are provided for making the test. In such cases some simple external instruments are indispensable for rapid and efficient fault-finding. Great care is necessary when checking transistors with test equipment which applies a d.c. voltage.

Multi-range test meter

This instrument usually consists of a sensitive moving coil meter arranged in a box with the necessary circuits and switching to read various values of a.c. and d.c. voltage and current. A battery or other low voltage source for measurement of resistance and, possibly, capacity is often included. It is usually provided with insulated terminals and a pair of long well-insulated leads terminated in test-prods. In use, the instrument is set by its selector switches to the required range of voltage, current, etc., and the prods applied to the points between which it is desired to make the measurement.

The best of these instruments are fitted with automatic safety cut-outs to protect them from damage due to overload; in all cases, however, it is desirable that the instrument should be set to the highest range expected in the test. Where the instrument is fitted with an OFF position it is good practice to return it to this position after each measurement, re-setting the switches for each operation as necessary. Where an OFF position is not provided the best practice is to leave the meter, between operations and when not in use, set to the *highest voltage* range, since accidental connection of it to any circuit is then least likely to cause damage to it (or to the circuit). A meter set up to read alternating voltage or current is liable to be damaged if it is connected to a direct voltage source. It is advisable to renew the prods immediately if the insulation becomes worn or cracked.

Insulation measurement—'The Wee Megger'

It is often required in a radar set to check the value of a high resistance, or to measure the insulation resistance of a cable or terminal. The ordinary multi-range meter rarely has a resistance range high enough to do this satisfactorily. Moreover, the voltage at the terminals of a multi-range meter in the 'resistance' position is extremely low, so that it provides no check of insulation (of condensers, &c.) under high-voltage conditions. An excellent instrument, measuring up to several megohms, is the 'Wee Megger'. This develops a voltage of 250 or 500 volts and is operated by turning a handle to rotate a small magneto generator. In use, its test prods are clipped across points between which the resistance is required to be known, the handle turned and the meter indication read on the scale provided.

The 'Wee Megger' develops a comparatively high voltage; if, therefore, the insulation of a condenser is checked it should first be ascertained that the rated working voltage of the condenser is greater than the voltage output of the 'Wee Megger'. Unpleasant shocks can, it should be noted, result from misuse of the instrument.

Measurements on components

It is important that before any component is checked for insulation, resistance, capacity, &c., its ends should be isolated from the rest of the circuit wiring. The presence of other components in the circuits may otherwise upset the indication given by the instruments.

The oscilloscope

One of the most useful instruments for checking the correct functioning of units of the radar is the oscilloscope, which displays a graphical representation of pulsing waveforms.

The oscilloscope incorporates a small cathode-ray tube which produces a horizontal trace, the duration of which is variable to suit different requirements. By connecting a probe between the oscilloscope and the particular point in a circuit under investigation, a pulse waveform can be displayed on trace. Pulses which may be examined in this manner include trigger and modulator pulses, radar time base and calibrator outputs and, if necessary, the returning echoes after video amplification.

The oscilloscope normally has a number of possible input sockets and controls and its operation is best understood by reading the manufacturer's operating booklet which is supplied with the oscilloscope.

The pulse shape obtained on the oscilloscope should be compared with waveforms shown in the radar handbook. However, it is not common for pulses in any particular radar equipment to have the 'ideal' shape shown in the handbook. A small departure from this ideal shape is not a serious matter; a large difference, however, indicates that something is amiss.

The use of the wave monitor

The best way to demonstrate the use of the equipment is to take a typical example of its use to examine a voltage pulse in a radar set. If the trigger voltage wave form which is applied to the modulator valve is to be checked, the vertical plates are connected by well-insulated prods between the grid of the modulator valve and earth. The wave monitor time base is set running and adjusted to a suitable duration. The pulse shape will then be shown on the trace and can be compared with that shown in the manufacturer's handbook.

It is not common for the pulses in any particular radar equipment to have the 'ideal' shape shown in the handbook. A small departure from this shape is not usually a serious matter; a large difference, however, often indicates that something is amiss.

Meter readings

Where manufacturers have built test meters into their equipments, they normally specify in their handbooks the readings which should be obtained in the various test positions. Sometimes these readings are quoted as lying between certain limits; in such a case a reading obtained outside these limits usually indicates a fault. Some manufacturers, however, give one figure for the readings to be expected, and it is rare for this figure to be precisely obtained in practice on any particular set. In such a case a good practice is to take all the meter readings when the set is known to be functioning correctly and use them as a standard by which to judge future indications. If, however, a fault occurs on the set before this has been done it is helpful to take all the meter readings and suspect that those which differ most from the manufacturer's quoted figure indicate the location of the fault. It is not possible to lay down in general what tolerances on manufacturers' figures are permissible in such cases, but it is probably fair to assume that a reading within 10 per cent. of that quoted indicates satisfactory performance.

Where manufacturers advise the use of a multi-range test meter external to the set, their typical readings are usually quoted as being obtained using a named multi-range instrument on a particular range. The use of a different type of instrument, or the use of the same instrument on a different range, may result in some differences between the readings obtained and those quoted. If the only meter available is of a different type from that named by the manufacturer, the procedure of noting all readings obtained when the set is known to be working correctly is advisable.

Appendix V

A SHORT GLOSSARY OF TERMS

Absorption. The dissipation of energy in the medium traversed, which is one cause of attenuation (q.v.).

Aerial. The part of a radio (or radar) transmitting system from which electromagnetic waves are radiated into space, or that part of a receiving system by means of which waves are collected.

Afterglow. The slowly decaying luminosity of a fluorescent screen when it has ceased to be excited by an electron beam. (See *Persistence*.)

Amplitude. The maximum value of an alternating voltage or current above or below the datum.

Amplitude modulation. The process of varying the amplitude of a radio frequency wave to enable it to perform a required function.

Antenna. See *Aerial*.

Aperture (of radar aerials). The effective area of the radiating surface in the plane normal to the principal direction of propagation. The horizontal width is often used as a measure of aperture.

Attenuation. Loss of power in a radio wave due to absorption, scattering and reflection by the medium through which it passes and by objects in its path.

Automatic acquisition. The acquisition by the computer in an automatic radar plotter of echoes for computation and display.

Beam-width. The angular width of a beam, measured between the directions in which the power intensity is a specified fraction of the maximum, e.g. $\frac{1}{2}$-power beam-width.

Bright display. A PPI display in which special consideration has been given to the need to view in the ambient light on the bridge without having recourse to a visor or special screening.

CPA. (Closest point of approach). The least distance at which two ships will pass if they remain on their present courses and speeds.

Clutter. Echoes from waves or precipitation, which appear in random fashion and may cover considerable areas of the PPI to the detriment of the picture. The terms *sea-return*, *wave-echoes* and *rain-echoes* are sometimes used.

Computation. The calculations necessary to solving the speed vector triangle of a maritime encounter.

Data rate. Usually expressed as the interval between successive observations of range and bearing (raw data). Also called *updating interval*, it may also be applied to the updating of computed data.

Decibel (db). A unit which expresses the ratio between two levels of power or voltage. (See Appendix III.)

Definition. A measure of the degree of detail on a radar display. For a display to have good definition, a combination of good resolution and focus is required.

Differentiating circuit. A circuit used to reduce the effect of rain and other clutter. In this application it is sometimes known as *fast-time constant* (FTC). It operates in such a manner as to display only the nearer edge of an echo which is extended in range, and thus to break up the patch of saturation which would otherwise be caused by clutter. It operates in all areas of the screen, as distinct from swept gain, which operates out to only a short distance from the centre. The principle may also be used in other parts of the equipment, such as the calibrator and trigger units.

Digital read-out. A numerical (alpha-numerical) display, usually of raw or computed target data.

Discontinuities. Breaks in the supply of data to or from the plot, usually irregular, which interfere with the normal data rate.

Discrimination (resolution). A measure of the ability of a radar set to display separately the echoes of two targets which are close together in range and/or bearing.

Duct. A stratum of atmosphere which confines within its limits the propagation of an abnormally large proportion of radiation.

Echo. In radar: (1) The radio-frequency energy returned to the aerial as a result of reflection or scattering from an object.

(2) The representation of (1) on a radar display. An echo is referred to variously as *echo pulse, return, signal,* &c.

A false echo is one whose position on the display does not indicate the correct range and/or bearing of the target. False echoes are subdivided into *indirect echoes, multiple echoes, side-lobe echoes* and *second-trace echoes.* (See Chapter 7.)

Envelope (modulation envelope). A graph defining the variations in amplitude of successive oscillations in an amplitude-modulated wave.

Fast-time constant (FTC). See *Differentiating circuit.*

Feather. See *Plume.*

Fluorescence. The property of emitting light as the immediate result of electronic bombardment.

Fly-back. The rapid return of the spot to its starting point at the end of the useful trace or sweep of a repeating time base.

Focusing. In cathode-ray tubes: concentrating the electron stream to produce a sharply defined spot on the screen.

Frequency modulation (f.m.). The process of varying the frequency of a carrier wave to enable it to perform a particular function.

Gain. The increase in power between the input and output of an apparatus. In a radar receiver the effect of a gain control is analogous to that of the volume control of an ordinary broadcast receiver.

Image-retaining panel. A sensitized screen which, placed in proximity with a radar CRT, accepts the radar picture and retains it long enough (3-4 minutes) for it to be photographed into a TV-format display, e.g., Situation Display. Retention of the echoes of moving targets allows echo trails to be seen.

Interference. In radar: (1) Intermittent paints on a radar screen due to an external radar or other transmitter.

(2) The coming together of direct and reflected rays, causing a series of points of maximum and minimum amplitude.

Interrogator. A pulse transmitter used exclusively for triggering a transponder such as a racon.

Lobe. The boundary of the volume inside which the field-strength (or power or voltage) of a radio wave is everywhere greater than a chosen value. It is usually depicted by drawing its sections in the horizontal and vertical planes.

Manual plot. A radar plot in which echo selection is made by the observer, transfer to the plot is either manual or with electronic assistance, and any graphical computation is manual.

Markers. Electronic spots or lines on the PPI usually placed to assist appraisal of the situation or to facilitate plotting or navigation.

Paint. The bright area left on a PPI display by the brightening of the trace due to echo pulses or other signals.

Parallax. An error, due to the height of the bearing graticule above the face of the PPI, which occurs when reading off the bearing of an echo unless the plane containing the graticule, the echo and the eye of the observer is at right-angles to the PPI face.

Perceptibility. A measure of the ease with which an echo can be observed.

Persistence. The quality of slow decay of the trace in a cathode-ray tube. (See *Afterglow*.)

Plume. The feather-shaped echo produced on a PPI by an echo-box performance monitor.

Polar diagram. A method of representing a variation of field strength with change of azimuth. The diagram is plotted radially from the centre of polar (circular) graph paper.

Polarization. Description of the plane in which the electrical axis of a radar wave lies. It may be horizontal, vertical or circular (rotating).

Rain-clutter control. See *Differentiating circuit*.

Relative motion. A form of display in which the position of own ship, i.e. trace origin, is fixed and the echoes of all other objects move relative to own ship.

Resolution. See *Discrimination*.

Responder. See *Transponder*.

Return. See *Echo*.

Scan. To explore a region by the (usually automatic) continuous variation of the direction of a beam. The scanner comprises the aerial and the mechanism which causes the beam to scan.

Scattering. Indiscriminate re-radiation of incident waves.

Screen. The face of a cathode-ray tube.

Sea-clutter control. A control to reduce the effect of sea clutter. In this book, the swept gain or sensitivity/time control methods is implied.

Second-trace echo. An echo which returns to the radar during the period of the trace succeeding that corresponding to the transmitted pulse which caused the echo. The target range must be greater than the distance corresponding to the pulse interval.

Sensitivity. A measure of the weakest signal which a receiver is able to detect.

Sensitivity/time control (STC). See *Swept gain*.

Shadow. The region in which, under normal propagation conditions, the field strength from a radar transmitter is reduced by some obstruction.

Side-lobe. A subsidiary and unwanted beam from a radar aerial. Side-lobes are usually on either side of the main beam and much weaker.

Signal. A term used to denote energy reaching the display via the aerial, e.g. signal pulse. It is used in the names of certain units which handle this energy, e.g. signal mixer.

Signal-to-noise ratio. The ratio, at any point of a circuit, of signal power (or voltage) to total circuit noise power (or voltage).

Sweep. See *Trace*.

Swept gain. A control to reduce the gain of the i.f. amplifier from the beginning of the trace (sweep) by a decreasing amount so that after a certain interval of time normal gain is restored. Also termed sensitivity/time control (STC).

Synchronism. Synchronism exists between two or more periodic activities when they maintain a constant relationship in time.

Trace (sweep). In radar: the movement of the spot of the cathode-ray tube as determined by the time base. The trace origin is the position at which the spot 'rests' between scans and represents the position of own ship.

Transmission line. Any conductor or system of conductors (e.g. a waveguide) used to carry electrical energy from its source to a particular destination.

Transponder. A unit which, in response to pulses received from a radar set or interrogator, transmits a pulse or sequence of pulses which can be recognized by the interrogating station.

True motion. A form of display in which own ship, i.e. the trace origin, moves across the PPI with her true course and speed (to the scale in use). All echoes of other objects follow suit.

Vector. A line expressing the movement of the object to whose echo it is attached. Its direction indicates the True or relative course and its length the distance run in the time selected and hence its speed.

Video frequencies. The frequencies of the signals which may be applied to a cathode-ray tube to produce a radar display (or television picture). They may range from 100 Hz to several MHz.

Appendix VI

DEPARTMENT OF TRADE MERCHANT SHIPPING NOTICE No. M.779

THE MERCHANT SHIPPING (RADAR) RULES 1976

Notice to Owners, Masters, Mates and Crews of Merchant Ships

Provision of Radar on Vessels of 1600 GRT and over

1. The Merchant Shipping (Radar) Rules 1976 (SI:302/76) came into force on 1 April 1976. They introduced requirements for the provision of a radar installation in ships of 1600 gross registered tons and above registered in the United Kingdom. Provision is also made in respect of the siting and serviceability of the radar installation.

Additional Facilities and Practices

2. The Rules are restricted to requirements that relate to the radar installation itself. It is recommended, in the interests of safety, that the following additional facilities and practices should be observed by ships which are fitted with a radar installation under the Rules:

   (a) every ship should be provided with facilities for plotting radar readings. These facilities should be on the bridge from which the ship is normally navigated, and, where practicable, in close proximity to the radar

   (b) a radar set when used as an aid to navigation or collision avoidance should be in charge of a qualified radar observer (but see (j) below). For the purpose of this Notice a person shall be deemed qualified to be a radar observer on board a British Ship registered in the United Kingdom provided he holds—

   (i) a Radar Observer's Certificate granted by the Department of Trade; or

   (ii) a certificate of attendance at a Department of Trade approved radar simulator course; or

   (iii) a certificate recognized by the Department of Trade as being equivalent to either of the certificates mentioned in paragraphs (i) and (ii).

   Note.—Details of Radar Observer and Simulator training are given in M.637.

   (c) Whilst the ship is at sea a radar watch should be kept by a qualified radar observer who can be assisted by unqualified personnel. The radar watch may be discontinued when navigating in good visibility clear of navigational hazards or when the master considers that conditions do not justify such a watch to assist the safe navigation of the ship.

   (d) 'Radar Watch' means observing displayed radar information, the frequency of observation being dependent upon the prevailing conditions.

   (e) the performance of the radar equipment should be checked before sailing and at least once every four hours whilst a radar watch is being maintained. This should be done by using the performance monitor where fitted (Rule 3(2) of the Rules).

(f) every ship when engaged on international voyages outside Home Trade limits should be provided with at least one officer or member of the crew qualified to carry out radar maintenance (but see (j) below). For the purposes of this Notice an officer or crew member shall be deemed qualified to carry out radar maintenance provided he holds:
  (i) a Radar Maintenance Certificate granted by the Department of Trade; or
  (ii) a certificate in radar maintenance recognized by the Department; or
  (iii) a certificate of proficiency in radar maintenance granted at the conclusion of a radar manufacturer's course which has been approved by the Department of Trade.
(g) in every ship a record should be maintained in a log book which may be the 'Deck Log Book' of the times at which radar watch is commenced and discontinued, or when the equipment becomes unserviceable and the person keeping the watch should also enter details of the performance checks mentioned in (e) above.
(h) Brief details of all maintenance work carried out on the radar equipment should be kept in the ship.
(j) In recommending that ships to which the Rules apply should carry qualified observers (paragraph 2(b) above) and at least one officer or crew member qualified to carry out radar maintenance (see 2(f) above) the Department recognizes that it may not be practicable to meet these recommendations until sufficient trained personnel are available and it is realized that a period of some four years may be necessary for this purpose.

Provision of Radar on Vessels under 1600 GRT

3. In addition it is desirable that all sea-going ships below 1600 gross registered tons registered in the United Kingdom should, where practicable, have regard to the provisions in the Rules and to the additional facilities and practices set out in this Notice.

Instructions to Surveyors

A publication, Instructions to Surveyors (Merchant Ship Radar Installation) which gives further information about the radar installations required by the Rules and the special tools and equipment required in accordance with Rule 12 of the Rules is in the course of preparation and will be published as soon as possible.

Department of Trade
Marine Division
London WC1V 6LP
September 1976

Appendix VII

DEPARTMENT OF TRADE MERCHANT SHIPPING NOTICE NO. M.784

THE USE OF RADAR

Notice to Shipowners, Shipmasters and Seamen

This notice supersedes Notice No. M.517

Collisions have been caused far too frequently by failure to make proper use of radar; by altering course on insufficient information and by maintaining too high a speed particularly when a close quarters situation is developing or is likely to develop. It cannot be emphasized too strongly that navigation in restricted visibility is difficult and great care is needed even though all the information which can be obtained from radar observation is available. Where continuous radar watchkeeping and plotting cannot be maintained even greater caution must be exercised.

Recommendations on the use of radar, agreed at the 1960 Safety of Life at Sea Conference, have been printed as an Annex to the international collision regulations. These are contained in Statutory Instrument 1965 no. 1525, The Collision Regulations (Ships and Seaplanes on the Water) and Signals of Distress (Ships) Order 1965. This Annex, besides giving some internationally agreed recommendations on the use of radar as an aid to avoiding collisions at sea, also clarified the interpretation of Rule 16 in the radar context. The present Notice consists of some notes which it is hoped will help mariners to obtain the utmost benefit from their radar equipment.

Clear Weather Practice

Whether or not radar training courses have been taken it is important that shipmasters and others using radar should gain and maintain experience in radar observation and appreciation by practice at sea in clear weather. In these conditions radar observations can be checked visually and misinterpretation of the radar display or false appreciation of the situation should not be potentially dangerous. Only by making and keeping themselves familiar with the process of systematic radar observation, and with the relationship between the radar information and the actual situation, will officers be able to deal rapidly and competently with the problems which will confront them in restricted visibility.

Interpretation

(a) It is essential for the observer to be aware of the current quality of performance of the radar set (which can be most easily ascertained by a performance monitor) and to take account of the possibility that small vessels, small icebergs and similar floating objects may escape detection.

(b) Echoes may be obscured by sea or rain clutter. Adjustment of control to suit circumstances will help, but will not completely remove this possibility.

(c) Masts and other obstructions may cause shadow sectors on the display. Notice No. M.535 on the fitting of radar sets makes provision for the measurement and recording of such sectors.

Plotting

To estimate the degree of risk of collision with another vessel it is necessary to forecast her nearest approach distance. Choice of appropriate avoiding action is facilitated by knowledge of the other vessel's course and speed, and one of the simplest methods of estimating these factors is by plotting. This involves knowledge of ship's course and distance run during the plotting interval.

Appreciation

(a) A single observation of the range and bearing of an echo can give no indication of the course and speed of a vessel in relation to one's own. To estimate this a succession of observations at known time intervals must be made.

(b) Estimation of the other ship's course and speed is only valid up to the time of the last observation and the situation must be kept constantly under review, for the other vessel, which may or may not be on radar watch, may alter her course or speed. Such alteration in course or speed will take time to become apparent to a radar observer.

(c) It should not be assumed that because the relative bearing is changing there is no risk of collision. Alteration of course by one's own ship will alter the relative bearing. A changing compass bearing is more to be relied upon. [See footnote*.] However, this has to be judged in relation to range, and even with a changing compass bearing a close quarters situation with risk of collision may develop.

Operation

(a) If weather conditions by day or night are such that visibility may deteriorate, the radar should be running, or on 'standby'. (This latter permits operation in less than one minute, whilst it normally takes up to five minutes to operate from switching on.) At night, in areas where fogbanks or small craft or unlighted obstructions such as icebergs are likely to be encountered, the radar set should be left permanently running. This is particularly important when there is any danger of occasional fogbanks, so that other vessels can be detected before entering the fogbank.

(b) The life of components, and hence the reliability of the radar set, will be far less affected by continuous running than by frequent switching on and off, so that in periods of uncertain visibility it is better to leave the radar either in full operation or on standby.

Radar Watchkeeping

In restricted visibility it is always best to have the radar set running and the display observed, the frequency of observation depending upon the prevailing circumstances, such as the speed of one's own ship and the type of craft or other floating object likely to be encountered.

---

*In the preliminary of Part D, Steering and Sailing Rules, of the Collision Regulations, (SI 1965 No. 1525) paragraph (2) reads:

'Risk of collision can, when circumstances permits, be ascertained by carefully watching the compass bearing of an approaching vessel. If the bearing does not appreciably change, such risk should be deemed to exist.'

Radar Training

It is essential for a radar observer to have sufficient knowledge and ability to recognize when the radar set he is using is unsatisfactory, giving poor performance or inaccurate information. This knowledge and ability can only be obtained by a full and proper training; experience alone or inadequate training can be dangerous and lead to collision or stranding through failure to detect the presence of other vessels or through misinterpretation of the radar picture.

Radar training courses have been established at a number of centres in the United Kingdom.

The Radar Observer Course is open to shipmasters, deck officers and intending deck officers of the Merchant Navy and those concerned with navigation in the Fishing Fleet. This course enables the mariner to obtain training in the operation and use of marine radar.

The Radar Simulator Course, open to shipmasters and senior deck officers, enables those officers to practise ship manoeuvring and collision avoidance on radar information. Considerable experience of realistic radar observation, interpretation and collision avoidance manoeuvres can be obtained during the five days of this course.

Information about these courses is included in Notice No. M.637, which can be obtained, together with a list of colleges at which the courses are held, from any Mercantile Marine Office.

Department of Trade (MNA 45/4/02)
Marine Division
London WC1V 6LP
November 1976

Editor's Note: Collision Regulations 1965 have been superseded by those dated 1972, and the Rule numbers should be amended accordingly.

Appendix VIII

SYMBOLS FOR CONTROLS ON MARINE NAVIGATIONAL RADAR EQUIPMENT

Extract from IMCO publication *Operational Performance Standards for Shipborne Navigational Equipment*.

The circles shown around symbols 4, 9, 10, 17, 22 and 23 are optional.

1.	⭕	OFF	TO IDENTIFY THE "OFF" POSITION OF THE CONTROL OR SWITCH
2.	⊙	RADAR ON	TO IDENTIFY THE "RADAR ON" POSITION OF THE SWITCH
3.	(symbol: broken circle with dot above gap)	RADAR STAND-BY	TO IDENTIFY THE "RADAR STAND-BY" POSITION OF THE SWITCH

4.		AERIAL ROTATING	TO IDENTIFY THE "AERIAL ROTATING" POSITION OF THE SWITCH
5.		NORTH UP PRESENTATION	TO IDENTIFY THE "NORTH UP" POSITION OF THE MODE OF PRESENTATION SWITCH
6.		SHIP'S HEAD UP PRESENTATION	TO IDENTIFY THE "SHIP'S HEAD UP" POSITION OF THE MODE OF PRESENTATION SWITCH

7.		HEADING MARKER ALIGNMENT	TO IDENTIFY THE "HEADING MARKER ALIGNMENT" CONTROL SWITCH
8.		RANGE SELECTOR	TO IDENTIFY THE RANGE SELECTION SWITCH
9.		SHORT PULSE	TO IDENTIFY THE "SHORT PULSE" POSITION OF THE PULSE LENGTH SELECTION SWITCH

10.	(pulse symbol in circle)	LONG PULSE	TO IDENTIFY THE "LONG PULSE" POSITION OF THE PULSE LENGTH SELECTION SWITCH
11.	(crescent symbol)	TUNING	TO IDENTIFY THE "TUNING" CONTROL
12.	(arc symbol)	GAIN	TO IDENTIFY THE "GAIN" CONTROL

#	Symbol	Name	Description
13.		ANTI-CLUTTER RAIN MINIMUM	TO IDENTIFY THE MINIMUM POSITION OF THE "ANTI-CLUTTER RAIN" CONTROL OR SWITCH
14.		ANTI-CLUTTER RAIN MAXIMUM	TO IDENTIFY THE MAXIMUM POSITION OF THE "ANTI-CLUTTER RAIN" CONTROL OR SWITCH
15.		ANTI-CLUTTER SEA MINIMUM	TO IDENTIFY THE MINIMUM POSITION OF THE "ANTI-CLUTTER SEA" CONTROL

16.		ANTI-CLUTTER SEA MAXIMUM	TO IDENTIFY THE MAXIMUM POSITION OF THE "ANTI-CLUTTER SEA" CONTROL
17.		SCALE ILLUMINATION	TO IDENTIFY THE MAXIMUM POSITION OF THE "SCALE ILLUMINATION" CONTROL OR SWITCH
18.		DISPLAY BRILLIANCE	TO IDENTIFY THE MAXIMUM POSITION OF THE "DISPLAY BRILLIANCE" CONTROL

19.		RANGE RINGS BRILLIANCE	TO IDENTIFY THE MAXIMUM POSITION OF THE "RANGE RINGS BRILLIANCE" CONTROL
20.		VARIABLE RANGE MARKER	TO IDENTIFY THE "VARIABLE RANGE MARKER" CONTROL
21.		BEARING MARKER	TO IDENTIFY THE "BEARING MARKER" CONTROL

22.	⬆ (up arrow in circle with trapezoid)	TRANS-MITTED POWER MONITOR	TO IDENTIFY THE ON POSITION OF THE "TRANS-MITTED POWER MONITOR" SWITCH
23.	⬆⬇ (up and down arrows in circle with trapezoid)	TRANSMIT/ RECEIVE MONITOR	TO IDENTIFY THE ON POSITION OF THE "TRANSMIT/ RECEIVE MONITOR" SWITCH

INDEX

Absorption, 305
  by atmosphere, 92
  by targets, 13, 74
Aerial (antenna), 305
  beam-width (relation to aperture and wavelength), 9, 25, 264
  Cheese, 24
  directivity, 9, 24
  gain, 9
  height, 70, 253
  height and radar horizon, 12, 146, 253
  height and vertical coverage diagram, 70, 78
  Slotted waveguide, 24, 214
  system, 24
  Tilted parabola, 24
Afterglow (persistence), 14, 305, 307
  Speckled, 34, 39, 49
  trails, 119, 189
Amplifier
  i.f. (intermediate frequency) description of, 36
  i.f. function of, 36
  i.f. main, 247
  i.f. pre- (head amplifier) 38, 246
  Logarithmic, 36
  Video, 38
Amplitude, 305
Anode (plate)
  Final, 40
Aperture, 305
Appraisal, aids, 277
Artificial line, see Delay line
Aspect (angle of view), 170
  of corner reflector, 197
  Importance to anti-collision, 170 171, 172, 264
  of land targets, 107, 148
  of ships, 112, 177
  of simple targets, 75
Atmosphere
  Absorption in, 93
  Non-standard, 89–92
  Standard, 88
Atmospheric discontinuities, 97, 124

Attenuation of radio wave, 305
  in atmosphere, 58, 92
  in fog, 93, 124
  in hail, 93, 95
  in rain, 93, 94
  in sandstorms, 93, 123
  in snow, 93, 95
Automatic acquisition, 305
Automatic frequency control (AFC), 50, 249
  tuning of klystron, 34
Automatic Relative Plot (ARP), 184, 265
Autoplot, 184
Azimuth
  Distortion in, 17, 153
  ring (bearing scale), 39
  synchronized circuits, 39, 42

Band-width in i.f. amplifier
  need for width, 38
Beam-width (beam angle), 61, 305
  Distortion due to, 17, 111, 112, 153
  effect on echo strength, 20
  effect on resolution, 20, 111–12
  effect on sea clutter, 121
  relation to aperture and wavelength, 9, 25, 264
  Vertical, 9, 20, 66, 293
Bearing
  accuracy, 293
  change of, 170–3
  cursor, 39, 52
  discrimination (resolution), 14, 112, 293
  distortion, 17, 153
  electronic cursor, 39, 44, 52, 166, 276, 278, 283
  errors, 53, 153, 154, 167
  measurement, 39, 52, 166
  measurement with racon, 213
  parallel index cursor, use of, 163, 164
  Relative, 43, 175, 176
  scale (ring), 39, 43, 52
  Steady, 171
Berthing
  in clear weather, 266
  in fog, 266
  VLCCs, 267

INDEX

Blind sector, 133, 253
Brightening pulse, 42
Brightness (brilliance)
    control, 14, 42
    effect on life of CRT, 49
    effect of storage on, 20
    range of variation on CRT, 38, 81, 98
    setting, change due to age of CRT, 46
    working adjustment of, 48
Buoy(s)
    echoes, 83, 112, 113, 159, 162, 289
    with radar reflectors, 159, 169, 197, 198, 202, 207

Calibration rings (markers), see Range rings
Calibrator circuit, 42
Cathode-ray tube (CRT), or Scope
    Action of, 14, 39
    Construction of, 39, 40
    Electromagnetic, 43
    life of, 49
Centring control, 53
Channel Navigation Information Service, 231
Chart, Radar, 168, 263
Closest point of approach (CPA), 273, 277, 305
Clutter, 36, 305
    Rain, 49, 122, 290
    Sea, comparison with radar buoy echoes, 202
    Sea, variation of strength with range, 84
    Sea (wave), 121, 290
        See also Sea-clutter control and Differentiating Circuit
Collision avoidance systems, 271–85
Collision risks
    alarm, 282
    assumptions about other ship, 187, 266
    avoiding action, 170, 173, 177, 194
    choice of range-scale, 187, 192, 193
    in coastal waters, 188
    indication of, 170, 174
    information needed, 172
    malpractices, 194
    in pilotage waters, 188
    radar for warning, 170
    responsibilities, 195
    when radar should be used, 187, 193
        See also Rule of the Road and Plotting

Compass
    conceded distance, 255
    Gyro, 43
    safe distance, 255
    stabilization of display, 43, 295
    Transmitting, 43
Computer-assisted plotting, see Plotting
Computing radar data, 278–85, 305
    methods of, 265
    See also Plotting
Controls
    Adjustment of, 46–55
    checking for correct operation, 245
    Operational, 46
    Pre-set, 46
    Unsynchronized, 40
Corner Reflector, 76, 197
Course
    Relative, 175, 177
    steered by other ship, 177
Coverage
    contour, 64
    diagram, 63
    diagram, effect of gain control, 65
    diagram, horizontal, 66
    diagram, vertical, 67, 76
Crystal
    AFC, 249
    AFC, current steady, 249
    AFC, current sweeping, 249
    mixer, Signal, 34
    Signal, 34
Cycle, 5

Danger from electric shock, radio-frequency and X-radiation, 242, 297
Data rate, 305
Decibel (db)
    formula and typical equivalents, 300, 305
Definition of Echoes, 305, 306
    factors affecting, 14
Deflection of spot
    Electromagnetic, 43
    Radial, 14, 39
Deflector coils
    function, 40
Delay line, 22, 23
Desiccators, 244
Detection, 191
Detection range
    Average, 124, 237
    effect of gain, 65

INDEX 325

Detection Range—*contd.*
 First, 71, 237
 increase by use of corner reflector, 113, 159, 162, 196, 197, 202, 207
 table, 286–91
 tables in Sailing Directions, 238
 of ship targets, 112
Detector
 Diode, 38
 First, 34
 Second, 38
Differentiating circuit (peaker), 36, 50, 306
 for rain-clutter control, 36, 50
Diffraction, 12
 effect on shadow sectors, 12
Digital read-out, 274, 281, 283, 306
Diode (valve)
 detector, 38
Discharge line, see Delay line
Discontinuities, 275, 306
Discrimination, see Resolution
Display, 13, 38, 98
 Analog display driver, 280
 A type, 113, 303
 adjustment, 48, 55
 'Bright', 132, 262, 276, 305
 centred, 14
 centring, 46, 52, 167
 Clearscan, 277
 of computed information, 265
 diameter, 98
 double stabilization, 43, 174
 Heading-upward, 43, 158
 interpretation, 98–120
 North-upward, 43, 158
 photographing and sketching, 107, 240
 PPI type, 14, 294
 Relative, 14, 119
 siting, 254
 Situation display, 277
 stabilization, 43, 295
 Television type, 222, 231, 269
 See also True motion
Distortion
 of echo in azimuth, 17, 112, 153
 of echo in range, 17, 112
 on PPI due to electrostatic charge, 143
 of reflector, 25
 of waveguide, 25
Duct, 90, 306
 Elevated, 91
Duplexing, see Transmit-receive

Echo (pip), 306
 False, 137, 290
 False, elimination of, 137, 253
 identification, 163, 196
 Indirect, 137, 290
 Indirect, in shadow sectors, 137
 Indirect (recognition), 141
 movement, 119, 164, 189
 Multiple, 131, 290
 paint, 14, 39, 306
 pulse, 14
 Ringing, 44
 Second-trace, 126, 291, 308
 Side-lobe, 142, 291
 trails, 119, 170, 189
Echo-box performance monitor
 for checking performance, 50, 52, 166, 236, 245, 297
 faulty, 245
 operation, 44, 52
 record of maximum reading, 236
 siting, 44, 255
 use for tuning, 50
Echoes from targets
 artificial structures, 85, 109, 288
 atmospheric discontinuities, 97, 124
 boats, 83, 111, 113, 207–8, 289
 bridges, 111, 289
 built-up areas, 85, 110, 288
 buoys, 83, 112, 113, 162, 289
 cable crossings, 111
 cliffs, 85, 99, 287
 clouds, 94, 122, 290
 corner reflectors, 76
 fog, 93, 124
 hail, 95, 123
 hills (mountains), 85, 287
  See also Ice
 isolated targets, 84, 111, 288
 land, distant, 99
 land, high, 85, 99
 land, low-lying, 85, 99, 287
 land/water boundary, 107
 overfalls, tiderips, &c., 109, 288
 rain, 94, 122, 290
 rollers, 121
 sandbanks, 85, 107, 109
 ships, 83, 112, 289
 snow, 95, 123, 290
 wake, 120, 189
 water spouts, 123
 waves, 84, 109, 290
Echoing Area
 Effective, 78
 Equivalent, 77, 197

Echoing Area—*contd.*
  Equivalent, of practical targets, 83, 207
Echoing characteristics, 13
Echo strength (power)
  aids to increasing, 196
  and angle of view, 74-6, 112, 148
  pulse-length and target slope, 73
  and range, 59, 80
  and resolution, 19, 261
  range of variation, 34, 38, 59
  Weakest detectable, 34, 36
  See also Detection range
Electron
  gun, 14
  stream, 14
Electronic bearing cursor, 39, 44, 52, 166, 276, 278, 283
  interscan, 221-2
Envelope, 306
Errors, 167, 189

Fast time constant (FTC) see Differentiating circuit and Rain
Faults in set
  Centre-spot wander, 52, 167
  Distortion, 143
  Hour-glass effect, 143
  Incorrect spacing of range rings, 167
  Index error, 168
  List of faults, 249
  Locating, 246-51, 302
  Non-linearity of time base, 167
  Sectoring, 143
  Serrated range rings, 143
  Spoking, 142
  Unequal spacing of range rings, 167
Field strength
  of radio wave, 57
  and range, 57
Flare, see Horn
Fly-back (of CRT spot), 39, 40, 306
Focus(ing), 306
  adjustment, 48, 49
  control, 40
  of radio waves, 9
  setting, 46
Fog
  Attenuation in, 93, 124
  Echoes from, 93
  effect on port operation, 219, 266, 268
Frequency, 5
  changing, 34
  Difference, 35
  drift, 35, 50

Frequency—*contd.*
  Intermediate (i.f.), 34
  modulation, 306
  of radar band, 294
  Radio (r.f.), 5, 13, 22
  spectrum, 7
  and wavelength, 5, 299
  Video, 308
Fuse
  failure, 246
  replacement, 246
  -wire table, 301

Gain, 306
  Aerial, 9
  for clearing screen, 108, 110, 114, 147
  control, 36, 40
  effect on coverage, 65
  effect on resolution, 36, 66, 154
  normal adjustment, 48
  for reducing clutter, 49, 122, 123
  for reducing side lobes, 66, 142, 168
  setting, and valve deterioration, 46
  in video amplifier, 38

Heading marker
  accuracy specified, 295
  contacts, 43, 243
  and echo paints, 43
  operation, 43, 53
  routine check of orientation, 43, 53, 54, 166, 245
Hertz (Hz), 5
Horizon
  distances, 299
  Geometrical, 12, 68, 299
  Optical, 12, 68, 299
  Radar, 12, 67, 68, 146, 299
Horn (also Hoghorn or Flare), 24, 243
Humidity
  effect on wave propagation, 89-90
  in non-standard atmosphere, 90
  in standard atmosphere, 88

Ice
  bergs, 83, 108, 287
  floes, 108
  growlers, 83, 108, 287
  pack ice, frozen, 108, 287
  re-frozen, 108
  sheet ice, 108, 287
Identification of targets
  aids to, 162, 196, 207, 230, 262
  distinguishing between targets, 113
  echo recognition table, 286-91

Identification of targets—*contd.*
  likely range of, 148, 237, 261
  at long range (landfall), 145, 261
  possible improvements, 262
Image-retaining panel, 275, 278, 306
Indicator, see Display and Cathode-ray tube
Interference, 307
  between direct and reflected rays, 68, 113, 207
  minima and maxima from reflection, 69, 114, 207, 216
  minima and maxima, lines on vertical coverage diagram, 70, 81
  from radar to other equipment, 255
  to radar from other radars, 132, 291
  to radar from radio, 132
Interscan, 221–2

Klystron (reflex) local oscillator
  AFC tuning, 34
  manual tuning, 35

Leading marks (ranges), 163, 263
Lighthouse
  echo strength, 85
Limit (-er-ing)
  in video amplifier, 38
Lobes, 307
  of vertical coverage diagram, 70
  See also Side lobes
Log, radar, 235, 238, 240
Logarithmic Amplifier, 36, 122
Look-out, 191
Luneberg lens, 201

Magnetic field
  in CRT, 43
  Earth's, 52
  in magnetron, 23
Magnetron, cavity
  first developed, 4
  operation in transmitter, 22, 23
Maintenance
  Simple, 242–51, 298
  Routine mechanical, 244
  Routine electrical, 244
    See also Faults, Performance checking and individual components by name
Manual/compass control, 53, 245
Markers, 276, 277, 284, 307
  See also Range

Meteorological conditions
  and ducting, 90
  effect of, 88–97
  and non-standard propagation, 91
  Standard, 88
Minima and Maxima (interference), 69, 81
Modulator, 22, 23
Monitor(-ing)
  of circuits for testing, 244, 303
  Wave, 303
    See also Echo-box performance monitor
Motor alternator, 22, 243

Navigation
  anchoring, 164
  berthing, 165, 266
  in buoyed channels, 162
  in coastal waters, 152, 188
  landfall identification and fixing, 144
  position on bridge, 254
  practice in clear weather, 144, 164, 255, 256
  preparatory action, 145, 147, 159, 165
  radar clearing lines, 154
  radar with other methods, 148
  radar range circles, 153
  radar range and radar bearing, 154
  radar range and visual bearing, 152
  radar in clear weather, 188
    See also Pilotage
Noise (electronic or thermal)
  level on PPI, 39
  and receiver sensitivity, 34

Orientation of display
  checking periodically, 43, 46, 53, 54
  heading-upward, 43, 158
  north-upward, 43, 158
  by picture-rotate control and heading marker, 43, 53, 55
  in relation to azimuth ring, 43
  stabilized Head-up, 173
Oscilloscope, 303

Parallax, 307
Peaker, see Differentiating circuit
Perceptibility (of echoes), 237, 307
Performance
  checking routine, 46, 55, 245
  monitor, specified requirements, 297
  specification, 260, 292–8
    See also Echo-box performance monitor

Photographic Radar plot (P.R.P.), 184, 231, 265
Picture-rotate control, 43, 53, 55
Pilotage
  collision warning, 188
  entering port, 219
  general considerations, 156, 158, 165
  by radar, 158
  visual, 159, 267
    See also Channel Navigation, and Shore-based radar
Pip, see Echo
Plan Position Indicator (PPI), 14, 262, 294
  ambient light level, 222
  off-centring, 221
  TV monitor display, 222, 231, 269
Plotting
  A.R.P., 184, 265
  automatic, 278–85
  Autoplot, 184
  CAS system, 282
  checking avoiding action, 180
  closest approach, 175, 177, 183, 185, 273, 277, 305
  compass datum, 176, 182
  computer-assisted, 224, 278–85
  Databridge, 279
  diagram, relative, 176
  Digiplot, 279
  electronic computers, 185, 260, 265, 278–85
  fully automatic, 172, 265, 278–85
  heading-upward plot, 176
  man-power needed, 259, 274
  manual, 274, 276, 284, 307
  mechanical aids to, 182–5
  methods, 173–85
  north-upward plot, 176
  other ship's course, 177
  Predictor, 279
  P.R.P., 184, 231, 265
  RAS plotter, 182
  Rate of approach, 171, 186
  reflection plotter, 184, 265, 276
  relative plot, 176–84, 189
  reporting from plot, 189
  selection of targets, 186, 265, 277–83
  systems, 275
  TM/CA, 184, 276
  true plot, 174
    See also Collision and Rule of the Road
Plume (from performance monitor), 307

Plume—*contd.*
  appearance when fault present, 245
  length, 52, 245, 255
Polar diagram (polar plot), 307
  Radiation, 60, 64, 66
  Target, 74, 83, 200
Polarization, 6, 25, 294, 307
Power supplies
  total consumption, 296
Presentation switch, 48
Propagation (of radio waves), 11, 57
  Anomalous (anoprop), 89, 124
  over curved Earth, 67
  with directional aerial, 63
  over Earth of increased radius, 67, 68, 71, 89
  far zone, 60, 81
  with non-directional aerial, 57
  under non-standard meteorological conditions, 89–92
  under standard meteorological conditions, 88
  variation of field strength with range, 57
Pulse
  D.c., 22
  Echo, 14, 38
  Radar, 9, 24
  R.f., 10, 13, 24
  Square, 10, 22
  switching, 19 (Table I)
  Synchronizing (sync) or Locking, 24
Pulse-forming network, see Delay line
Pulse-length (duration, width), 10
  effect on minimum range, 10
  effect on resolution (discrimination), 111, 112
  effect on storage, 20
  on PPI, 14
Pulse-repetition frequency (p.r.f.) or Pulse-repetition rate
  effect on storage, 20
  and maximum range, 10
  normal value, 9
  and second-trace echoes, 125

Racon, 197, 210–18
  bearing measurement, 213
  coding (identification), 212
  comparison with ramark, 215
  cross-band, 210
  in-band, 210, 212
  maximum detection range, 213
  minima due to interference, 216

Racon—*contd.*
  principle of operation, 210
  swept-frequency, 211
Radar absorbent material, 141, 254
Radar beacon
  cross-band and in-band, 209–10
  interference, 215, 216
    See also Racon and Ramark
Radar cross section, see Echoing area
Radar reference lines, 169
Radar reflectors, 76, 197
  collapsible, 208
Radar reliability, 242, 260
Radar set
  at short notice, 256
  basic characteristics, 5, 13, 19
  basic components, 22
Radar Simulator courses, 256
Radiation diagram (pattern), 60, 64, 66
Radio waves
  behaviour, 8
  bending of, 12, 67
  nature of, 5
  polarization, 6, 25, 307
  velocity, speed of, 5, 299
Rain
  Attenuation due to, 94
  clutter, 122, 290
  echoes, 94, 122, 290
  effect of anti-clutter controls, 123
  effect of circular polarization, 25, 307
  effect of differentiating circuit, 36, 49
  effect of logarithmic amplifier, 36
  rainfall statistics, 96
Ramark(s), 197
  cross-band, 209
  coding (identification), 215
  as leading mark, 263
  in-band, 214, 215
  minima due to interference, 216, 218
  principle of operation, 214
  and racon, 209, 215, 218
Range
  accuracy, 46, 54, 293
  circles, 153
  discrimination (resolution), 16, 111
  distortion, 17
  errors, 54, 167
  marker, interscan, 221
  marker, variable, 42, 54, 165
  Maximum, 12

Range—*contd.*
  measurement on PPI, 54, 165, 222
  Minimum, 10, 192
  rings, 42, 54, 166, 250, 293
  rings, serrated, 143
  scale, 42
  scale, specified requirements, 293
  switch (selector), 40
Receiver (radar)
  for cross-band ramarks and racons, 210
  principal components, 26–45
  sensitivity, 36, 65
  tuning, 50
Rectification
  in receiver (detection), 36
  by simple diode, 38
Reflection (of radar ray)
  from cone, 77
  from corner reflector, 76, 197
  from cylinder, 77, 84
  Diffuse (scattering), 74, 77
  equivalent to re-radiation, 13, 74
  from flat plate (plane), 75
  from perpendicular planes, 76
  Specular, 75
  from sphere, 74
    See also Echoes from targets
Reflection plotter, 184, 265, 276
Reflector
  aperture, 305
  Radar, 197
  size and relation to wavelength and beam-width, 9, 24, 264
Reflector (corner), 76, 197–209, 262
  on boat's mast, 208, 262
  on buoy, 113, 198, 202, 207, 289
  collapsible, 208
  dihedral, 76, 197
  Luneberg lens, 201
  need for identification system, 207
  new materials, 201
  octahedral cluster, 198
  pentagonal cluster, 198
  size and equivalent echoing area, 206–7
  size to overcome wave echoes, 202
  trihedral, 197
Refraction
  Atmospheric, 67
  Standard, 11, 88
  Sub-, 89, 125
  Super-, 90, 125
    See also Meteorological conditions

Refractive index
  rate of change, 11, 88, 92
Relative motion, 307
Resolution (discrimination), 306
  in bearing, 16, 111, 293
  effect of differentiation, 36
  effect of gain, 36, 65
  improvement in future, 262, 264
  Minimum area of, 17, 112, 264
  in range, 16, 111, 293
  relationship with echo strength, 20
  spot size limitation, 16, 111
Resonant cavity
  in echo box, 44
  in klystron (resonator), 34
  for tuning radar frequencies, 35
Response of targets, 13, 72, 78, 83
River radar, 21
Rule of the Road, 191–5
  Ascertaining the position of other ship, 193
  Court judgements, 194
  Radar introduced, 191
  Safe speed, 192
  Sound signal, 192
  See also Collision and Plotting

Safety Measures, 242, 254, 297
Safety switches, 242, 244
Saturation
  of screen, 36, 49
Scanner, 9, 307
  housing, 25
  motor, 242
  maintenance, 243, 249
  rotation speed, 294
  siting, 253
  See also Aerial and Reflector
Scattering, 13, 307
  See also Reflection of radar ray
Scope, see Cathode-ray tube and Display
Screen (CRT)
  fluorescent or sensitized, 14, 308
  saturation of, 36, 49
Sea clutter
  and corner reflectors, 197, 202
  by multiple trace, 122
  nature of, 121
  reduction of, 121
Sea-clutter control (suppressor), 36, 42, 308
  swept gain, adjustment of, 36, 49, 84, 121, 202

Sea-clutter control—*contd.*
  swept gain, danger of losing weak echoes, 36, 49, 84
Sensitivity
  Receiver, 36, 307
Sensitivity-time control (STC), see Sea-clutter control
Setting-up procedure, 54
Shadow, 308
  area, 132, 153
  Radar, 99
  sectors, 132, 136, 240, 253
Ship-shape radar, 21, 265
Shore-based radar
  functions, 219
  future development, 269
  Liverpool installation, 223
  operational procedure, 230
  radio link (PPI) to ship, 269
  ship-shore communication, 219, 220, 224
  Teesport installation, 222, 231, 269
Side lobe(s)
  Aerial design to reduce, 24, 61
  of aerial radiation diagram, 61, 142, 308
  echoes, 142, 291
  echoes, use of gain control to reduce, 65, 142, 168
  of target diagram, 75
Signal mixer, 34
Siting
  Aerial, 253
  Display console, 254
  Echo box, 44, 255
  Scanner, 253
  Transmitter, 254
Small craft, 165
Solid-state devices, 23, 261
Sound signals, 192
Speckled background (on PPI), 34, 42, 49
Speed
  of approach or closing, 171, 186
  of deflection (spot), 14, 40
  of light waves, 8
  of radio waves, 8, 299
  of rotation of aerial/scanner, 294
  Safe (Rule of the Road), 192
  of sound waves in air, 1
  -vector diagram, 176–85
Spot (CRT), 14
  centre-spot wander, 52, 167
  centring, 53, 54
  fly-back, 40

INDEX

Spot—*contd.*
  radial deflection, 14, 39
  size (width) of, 16, 111, 264
  speed of movement, 14, 16, 40
Stabilization
  of display, 43, 158, 167, 173, 295
Static power converter, 243
Storage effect (of echoes on PPI), 20
Suppression, } see Sea-clutter
Suppressor } control
Sweep, see Trace
Swept gain, see Sea-clutter control
Symbols for controls, 314–21
Synchronized circuits
  Azimuth-, 39, 42
  Time-, 39, 40
Synchronizing (sync) pulse, 24

Targets
  aspect (angle of view), 13, 75, 107, 112
  equivalent echoing area, 78
  height, 80, 82
  movement of, appreciation, 114, 165, 188, 277–83
  polar diagram, 74, 75, 83, 200
  poor targets, 83
  probable detection ranges, 286–91
  projected area, 74, 77, 78
  response of practical targets, 83–7
  response of simple targets, 13, 72, 78–83
  shape, 13, 73, 74
  size, 13, 72
  surface composition (texture), 13, 72, 78–83
  vertical extent of, 73, 78, 85
    See also Echoes from, and Identification
TR cell, 25
  Broad band, 25
Temperature (atmospheric)
  change in non-standard atmosphere, 89
  change in standard atmosphere, 88
  inversion, 91
Test equipment and meters, 257, 302
  See also Monitoring
Thyratron, 23
Time-base, 40
  linearity, 14, 40
  non-linearity, 167

Trace (sweep), 308
  formation, 14, 39
  linearity, 14, 40–2
  origin, 39, 307
  rotation, 14, 39, 43
  rotation speed, effect on storage, 20
  rotation speed, practical values, 20, 295
  rotation system, maintenance, 244, 249
Transformer
  Oil-filled, checking, 244
Transistors
  care when testing, 246, 252, 257, 302
  replacing valves, 261
Transmit receive (TR) or Duplexing
  arrangements, 25, 26
  separate aerials, 25
Transmitter
  general requirements, 13, 22
  principal components in outline, 22
  siting, 254
Transporter, 308
Trapping, see Duct
Trigatron, 23
Trigger
  circuit, 22, 23
Trouble shooting, see Faults
True motion
  controls, 48, 50, 55
  echo trails, 119
  general, 39, 308
  pilotage, 158, 164
Tube (vacuum tube), see Valve
Tuning
  AFC control, 50
  drift, 50
  manual control, 50
  meter, 50, 55
  by performance monitor, 50, 55
  on sea clutter, 50, 55

Valve (vacuum tube)
  detector, 39
  deterioration, 246
  life, 246
  phasing out, 261
  rectifier, 36–8
  See also Diode
Variable range marker (strobe)
  accuracy, 54
  calibration, 42, 54, 166
  circuit, 293

Variable range marker—*contd.*
  control, 244, 276
  method of use, 54
Vector, 275, 308
Video frequencies, 308

Wave(s)
  Electromagnetic, 5, 6, 8
  Light, 6, 8, 11, 12, 67
  monitor, 303
  motion, 5
  Sound, 1
  in water, 5, 68
    See also Radio waves

Waveguide, 24
  effect of distorting, 24
  losses in, 25
  moisture in, 24
  slotted, 24
Wavelength
  beam-width and aerial aperture, 9, 25, 264
  and bending, 8, 11
  and frequency, 5, 299
  of marine radar, 8, 294
  of usable radio waves, 7
  and vertical coverage diagram, 70
'Wee Megger', 302